Revise AS Chemistry for /

Paddy Gannon

Heinemann Educational Publishers
Halley Court, Jordan Hill, Oxford, OX2 8EJ
Part of Harcout Education

Heinemann is the registered trademark of Harcourt Education Limited

© Paddy Gannon, 2000

Copyright notice

All rights reserved. No part of this publication may be reproduced in any material form (including photocopying or storing it in any medium by electronic means and whether or not transiently or incidentally to some other use of this publication) without the prior written permission of the copyright owner, except in accordance with the provisions of the Copyright, Designs and Patents Act 1988 or under the terms of a licence issued by the Copyright Licensing Agency Ltd, 90 Tottenham Court Road, London W1P 0LP. Applications for the copyright owner's written permission to reproduce any part of this publication should be addressed to the publisher.

First published 2001

10-digit ISBN 0 435583 18 2
13-digit ISBN 978 0 435583 18 7

10 09 08 07 06
10 9 8 7 6 5 4 3

Development Editor Paddy Gannon

Edited by Peter Millward

Index compiled by Diana Boatman

Designed and typeset by Saxon Graphics Ltd

Printed and bound in Great Britain by Ashford Colour Press Ltd, Gosport, Hants

Tel: 01865 888058 www.heinemann.co.uk

Contents

Introduction	iv
Chemistry for AQA – Assessment	v

AS Module 1 – Atomic structure, bonding and periodicity — 2

Atomic structure
Fundamental particles	2
Mass spectrometry	4
Electronic arrangement	6
Electronic configuration	8
Ionisation energy	10

Amount of substance
Using moles	12
Empirical formula	14
Reacting amounts	16
Reacting volume calculations	18
Molarity and concentration	20

Bonding
Bondage	22
Bond polarity	24
Intermolecular forces	26
States of matter	28
Structures	30
Shapes of molecules	32

Periodicity
Periodicity	34
Period 3 – sodium to argon	36
Group II – alkaline earth metals (I)	38
Group II – alkaline earth metals (II)	40
Exam style questions	42

AS Module 2 – Foundation physical and inorganic chemistry — 44

Energetics
Enthalpy change	44
Hess's law	46
Bond enthalpies	48

Kinetics
Rate of reaction	50

Equilibria
Equilibria	52
Industrial processes	54

Redox reactions
Redox	56
Redox reactions	59

Group VII – the halogens
Group VII – the halogens	62
Halide ions	64
Chlorine	66

Extraction of metals
Reduction of metal oxides	68
Other reduction methods	70
Exam style questions	72

AS Module 3 – Introduction to organic chemistry — 74

Nomenclature and isomerism
Naming organic compounds	74
Isomerism	77

Petroleum and alkanes
Alkanes (I)	80
Alkanes (II)	82

Alkenes and epoxyethane
Alkenes	84
Reactions of alkenes (I)	86
Reactions of alkenes (II)	88

Haloalkanes
Haloalkanes	90

Alcohols
Alcohols	92
Exam style questions	95

Answers	97
Index	106

Introduction

This book is designed to help you study the AQA Advanced Subsidiary General Certificate of Education in Chemistry (Specification 5421). It concentrates on the facts you need to know to understand the concepts and provides you with opportunities to practise answering questions.

The content of each module follows the specification and divides it up into manageable chunks.

Each chunk is covered by a number of pages in which you'll see:

- the specification material **condensed** and **summarised** into the most important points and squeezed unbelievably into 112 pages;
- **tips boxes**: these point out things you need to watch out for or give you general tips about how to get the Chemistry right;
- **diagrams**: these are drawn simply, so you can reproduce them in exams, if needed;
- **quick check questions**: these check that you're understanding the material covered on a particular double page spread.

At the end of each module there are **exam style questions**: these dig a little deeper than the quick check questions and give you an idea of the things AQA may ask you in exams. At the end of the book you will also find brief **answers**: it's always nice to double check you are heading in the right direction.

Do you speak Chemistry?

This may seem like a silly question. But one of the hardest things in Advanced Subsidiary (AS) Chemistry is getting to grips with a whole new language – the *language of Chemistry*. Without this you'll be lost and unable to communicate any understanding that you have.

Key words are highlighted in **bold**, so you can immediately see that a new or important word has been introduced or used. Make sure that you learn the meaning of these words, as you are going to need them when it comes to those horrid module tests and exams.

So why go through this and study Chemistry?

Chemistry gives you the knowledge that allows you to see and understand the world in a special way. It is both **essential** and **fascinating**. You also get to wear Clark Kent style safety glasses; you'll know how to completely dissolve the bodies of your enemies; you'll get permanent goggle marks (which is cheaper than a tattoo) and you'll be able to say the word 'buckyballs' without bursting into a fit of laughter. What more could you want?

And remember...

>...Chemistry is pHun... ☺

So have pHun....

Paddy Gannon
The Lakes

Chemistry for AQA – Assessment

AS Chemistry

The scheme of assessment for **AQA AS GCE in Chemistry (5421)** involves three exams to cover the three modules you'll study, and some lovely practical assessment.

AS Assessment units	Type of exam	Length (h)	Content of test	% of AS (% of total A level)
1	Structured questions	1.5	Module 1	30 (15)
2	Structured questions	1.5	Module 2	30 (15)
3a	Structured questions	1.25	Module 3	25 (12.5)
3b	Centre-assessed coursework (or Practical exam)	Varies in length (2)	Practical chemistry	15 (7.5)

A2 Chemistry

If you go on to study for the full **Advanced level** (AS + A2) qualification you must take the second part, which involves studying for the **A2 modules 4, 5 and 6.** You then sit the appropriate assessment units.

The assessment for AQA A2 Chemistry (6421) involves three more exams and some more lovely practical assessment. The marks from these units of assessment are added to the AS assessment units with the appropriate weighting shown below.

A2 Assessment units	Type of exam	Length (h)	Content of test	% of total A level*
4	Structured questions	1.5	Module 4	15
5	Structured questions	2	Module 5 (some of 4 and AS)	20
6a	Synoptic objective questions	1	All modules AS and A2	10
6b	Centre-assessed coursework (or Practical exam)	Varies in length (2)	Practical chemistry	5

* Total A level = AS + A2

In both AS and A2 longer answers will be marked for the quality of written language used.

Synoptic assessment

The questions AQA will ask in A2 Assessment Unit 5 will be on Module 5 and some material from Module 4. These questions can overlap with AS material, as A2 builds on AS material.

A2 assessment Unit 6a will draw together knowledge, understanding and skills learned in all the AS and A2 modules.

Re-sits

You can re-sit each assessment unit once only. If you do need to re-sit, your better result will count.

Module 1: Atomic structure, bonding and periodicity

Fundamental particles

We are pretty sure that all matter in the Universe is made up from 92 naturally occurring **elements**. These are found listed in the **Periodic Table**. Each atom of an element contains a certain number of sub-atomic particles which govern the element's **physical and chemical properties**.

Sub-atomic particles

- There are three sub-atomic particles you must known about, **protons**, **neutrons** and **electrons**.
- Protons and neutrons are found in the **nucleus** and the electrons **orbit** the nucleus.
- Their actual masses are very small.

> There are other sub-atomic particles but we let Physicists worry about those.

Relative mass

Since the actual masses of sub-atomic particles are very small and difficult to deal with, we usually use **relative mass** to describe them.

- The proton is assigned a relative mass of **1**.
- The masses of the other sub-atomic particles are described relative to (or compared to) the mass of one proton.

Relative charge

- The protons and electrons have an equal but opposite charge of +1 and −1 respectively.
- The neutrons do not have a charge. Unsurprisingly, they are **neutral**.

Particle	Mass	Charge	Position
proton	1	+1	in nucleus
electron	1/1836	−1	outside nucle
neutron	1	0	in nucleus

Atomic number

- The atomic number of an atom is **the number of protons** in the nucleus.
- It is given the letter **Z** in calculations.
- The number of protons in an atom tells us the **element** to which the atom belongs.
- It is shown as a subscript (bottom no.) before the symbol of the element.
- The atomic number is also known as the **proton number**.

> If you're asked 'What is the relative charge of a proton?' Always state '+1' or 'plus one'... don't just say 'positive' – you'll lose marks.

Mass number

- The **mass number** is the sum of the numbers of protons and neutrons in the nucleus.
- The mass number is always shown as a superscript (top no.), **A**, before the symbol of the element.
- Some books call mass number the **nucleon number**.

Facts you should know about atomic and mass numbers

You need these numbers to work out the number of sub-atomic particles in atoms up to krypton in the periodic table.

- The number of **protons** in an atom is equal to the **atomic number** → Z.
- Since atoms are neutral, the number of **electrons** will be the same as the number of **protons** → Z.
- The number of **neutrons** is the difference between the **mass number** and the **atomic number** → $(A-Z)$.

For example: a lithium atom

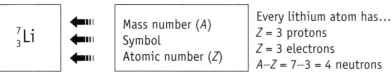

Every lithium atom has...
$Z = 3$ protons
$Z = 3$ electrons
$A-Z = 7-3 = 4$ neutrons

Isotopes

- **Isotopes** are atoms of an element with the same atomic number but different mass numbers.
- They have the same number of protons and electrons, but **different numbers of neutrons**.
- They therefore have **different masses**.
- They can be **natural** or **artificial**.

For example the element chlorine has two isotopes:

$$^{37}_{17}Cl \qquad ^{35}_{17}Cl$$

Each chlorine atom has: 17 protons   17 protons
                          17 electrons   17 electrons
                          20 neutrons   18 neutrons

> 'Iso' means 'the same' or 'equal'...like isobar – a region of equal pressure...

✓ Quick check 1, 2

Properties of isotopes

- Isotopes have the same **chemical properties** since they have the same number and arrangement of electrons and it is the electrons that dictate the chemical properties of the elements.
- Isotopes have slightly different **physical properties** since they have different numbers of neutrons and it is mainly the neutrons (and protons) that determine properties like the **mass** of the atom.

The lighter isotopes of an element will have...
1 **lower** densities
2 **faster** rates of diffusion
3 **lower** melting and boiling temperatures
4 different peaks with different m/z values on a mass spectrum.

▶▶ *There is more about mass spectra on page 4.*

Other important isotopes: Hydrogen
The isotopes of hydrogen have **different names**:

	hydrogen (H)	deuterium (D)	tritium (T)
	^1_1H	^2_1D	^3_1T
Each atom contains:	1 proton	1 proton	1 proton
	1 electron	1 electron	1 electron
	0 neutrons	1 neutron	2 neutrons

> Hydrogen is unusual because it's the only isotope of an element without neutrons.

✓ Quick check 3, 4

Quick check questions

1. Where is most of the mass of an atom concentrated?
2. Give definitions for the following terms:
   a atomic number, **b** mass number, **c** isotope,
3. An isotope has six protons, six electrons and eight neutrons. Name the element.
4. Work out the number of protons, electrons and neutrons in the following:
   a $^{12}_{6}C$, **b** $^{40}_{20}Ca$, **c** $^{11}_{5}B$, **d** $^{39}_{19}K$, **e** $^{27}_{13}Al$.

Module 1: Atomic structure, bonding and periodicity

Mass Spectrometry

Isotopes are identified using a technique called **mass spectrometry**. It's used to determine the **relative atomic mass** of an element and the **relative molecular mass** of a compound. It's also used to calculate the relative abundance of isotopes in a sample – these are calculated from mass spectra.

How it works

There are five processes going on in the mass spectrometer that you must know about. They are listed below and shown in the diagram.

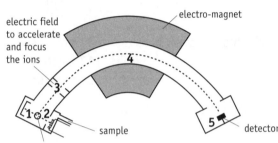

1 **Vaporisation** – turns the sample into a gas.
- The sample is **vaporised** by heating it electrically.
- It is then passed into the **ionisation chamber**.

2 **Ionisation** – turns the gas into ions.
- The sample is then bombarded with **high-energy electrons**.
- This causes it to lose electrons and ionise, producing **unipositive ions**.

3 **Acceleration** – gets all the ions moving at the same speed.
- An **electric field** attracts the ions and accelerates them towards the detector.
- They all leave the ionisation chamber with similar **velocities**.
- This **ion stream** contains ions of **different isotopic masses**.

4 **Deflection** – separates the ions by mass.
- The ion stream travels through a **magnetic field**.
- The ions are **deflected** by the field.
- Heavier ions have more **momentum** than lighter ions so are deflected less.
- This sorts out the mixture of ions in the ion stream into separate streams of ions with single **isotopic masses**.

5 **Detection** – identifies the mass and relative amount of each isotope.
- As the magnetic field is gradually increased, the streams of isotopes hit the detector one at a time *(the other isotopes simply hit the wall of the instrument)*.
- The detector converts the information into a **mass spectrum**.
- The height of each peak is **proportional** to the **relative abundance** of the isotope.

✓ *Quick check 1*

There is more deflection...
- the **lower** the **mass** of the positive ions;
- the **lower** the **velocity** of the positive ions;
- the **higher** the **charge** on the positive ions;
- the **larger** the **strength** of the magnetic field.

Determining relative atomic mass

The relative atomic or molecular mass of a species can be determined using the mass spectrometer.

If the ions are travelling...

- at the same velocity
- with the same charge and
- through the same magnetic field

...then the amount of deflection depends only on the **mass** of the ion.

> Species means the thing you are talking about, like an atom or an ion and only mass spectra for 'mononuclear' ions are usually asked about – these are ions with one atom.

For a certain magnetic field strength, ions will have a particular **mass (m) to charge (z) ratio** (m/z) value. But if the peaks in the spectrum are for **unipositive ions** ($z = 1$), then the m/z value is the same as the **mass number** of the ion.

For samples of an element each isotope present produces a **peak** and so a calculation is needed to work out the relative mass of the element.

Each isotope has a different mass and so each isotope will produce a peak with a different m/z value. Use the equation below to calculate the relative atomic mass.

> When you look at mass spectra the y axis may be labelled 'relative abundance' or '% abundance'...just think of these as being the same thing.

Relative atomic mass of element = $\dfrac{\Sigma(\text{isotopic relative masses } (m/z) \times \text{relative abundance (\%)})}{\Sigma(\text{relative abundances})}$

(where Σ = sum of)

Worked example

Calculate the relative atomic mass of chlorine from the mass spectra opposite. There are two isotopes of chlorine: Cl-35 with a relative abundance of 100.0% and Cl-37 with a relative abundance of 32.5%.

$$\text{Relative atomic mass of chlorine} = \frac{(35 \times 100.0) + (37 \times 32.5)}{(100.0 + 32.5)} = \frac{3500 + 1202.5}{132.5}$$

$$= \frac{4702.5}{132.5} = 35.491$$

∴ chlorine is given a relative atomic mass of 35.5 (To 3sf)

> If there's no scale you have to measure height with a ruler

✓ Quick check 2, 3

Quick check questions

1. List the five processes an isotope must undergo in the mass spectrometer.
2. Explain why the relative atomic masses for some elements are not whole numbers.
3. A sample of neon was found to contain three isotopes with the following relative abundances: 90.9% ^{20}Ne, 0.26% ^{21}Ne and 8.8% ^{22}Ne. Calculate the relative atomic mass of neon.

> Some questions make the total of the relative abundance more than 100. It's still a relative scale – tricky, but don't worry about it. Also remember that molecular mass can be asked about, not just atomic mass.

Electronic Arrangement

The electronic structure of an atom is very important as it controls the chemical properties of the element. When you studied Chemistry (or Science) at GCSE, you used the Bohr model of the atom to picture the structure and arrangement of electrons around the nucleus. At AS level, you study this model in more detail.

Orbitals, sub-levels and principal energy levels

An **orbital** is a region where it is highly likely an electron will be found.

- They are **negative charge clouds**.
- They can hold up to two electrons (each with opposite spin).

An electron **sub-level** (or sub-shell) is made up of one or more orbitals.

- The sub-levels are known as **s, p** and **d**.
- s, p and d sub-levels can hold **different numbers** of electrons.

A **principal energy level** (or shell) is a collection of sub-levels.

- The first principal energy level ($n = 1$) is **nearest** the nucleus.
- Each successive energy level is at a **higher** energy.
- All electrons around the nucleus will be in one of these energy levels.

✓ *Quick check 1*

The table shows the sub-levels and the maximum number of electrons in each principal energy level.

Principal energy level (n)	1	2		3			4			
Number of sub-levels (sub-shells)	1	2		3			4			
Name of sub-level (sub-shell)	s	s	p	s	p	d	s	p	d	f
Number of orbitals	1	1	3	1	3	5	1	3	5	7
Maximum no. of electrons in the sub-level	2	2	6	2	6	10	2	6	10	14
Maximum no. of electrons in the energy level	2	8		18			32			

> The maximum no. of electrons found in each shell = $2n^2$ (where n = principal energy level). E.g. if $n = 2$, $2 \times 2^2 = 8$

The shapes of the orbitals

- There is one s orbital in an s sub-level and it is spherical.
- There are three p orbitals in a p sub-level and they are dumb-bell shaped.
- There are five d orbitals in a d sub-level and they are a variety of complex shapes.

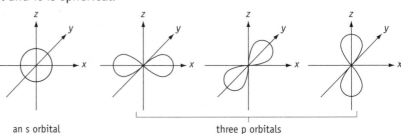

an s orbital   three p orbitals

Atomic structure

Rules for filling the principal energy levels

- Electrons always occupy the **lowest** energy sub-level first.
- The order is 1s 2s 2p 3s 3p **4s** 3d 4p for the elements hydrogen to krypton.
- This building up of electronic structure is called the **Aufbau principle**.
- An orbital holds 2 electrons, but electrons occupy the orbitals first as **unpaired electrons** rather than **spin-paired electrons** (they pair up after the sub-level is half full).
- This is known as **Hund's Rule**.

Use this diagram to help you remember the Aufbau principle ▶

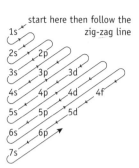

Oxygen's electronic configuration is $1s^2 2s^2 2p^4$

Principal energy level, sub-level, electrons

✓ Quick check 2

The filling of the orbitals

Using the above rules the principal energy level and sub-levels are filled.

The table shows the electronic configuration of the elements H ⟶ Kr

Group	I	II	III	IV	V	VI	VII	0
Period	ns^1	ns^2	$ns^2 np^1$	$ns^2 np^2$	$ns^2 np^3$	$ns^2 np^4$	$ns^2 np^5$	$ns^2 np^6$
1	H* $1s^1$							He $1s^2$
2	Li $2s^1$	Be $2s^2$	B $2s^2 2p^1$	C $2s^2 2p^2$	N $2s^2 2p^3$	O $2s^2 2p^4$	F $2s^2 2p^5$	Ne $2s^2 2p^6$
3	Na $3s^1$	Mg $3s^2$	Al $3s^2 3p^1$	Si $3s^2 3p^2$	P $3s^2 3p^3$	S $3s^2 3p^4$	Cl $3s^2 3p^5$	Ar $3s^2 3p^6$
4	K $4s^1$	Ca $4s^2$	Ga $4s^2 4p^1$	Ge $4s^2 4p^2$	As $4s^2 4p^3$	Se $4s^2 4p^4$	Br $4s^2 4p^5$	Kr $4s^2 4p^6$

s block (Groups I–II) ; p block (Groups III–0)

* hydrogen has an electronic structure similar to group I elements which is why it is put here
n = number of period

✓ Quick check 3

Quick check questions

1. What is a principal energy level?
2. State the Aufbau principle and Hund's rule.
3. Give the electronic configuration of Al and Ca.

Electronic configuration

To get the hang of how electrons fill up the energy levels and how electrons are arranged, you need to practice writing out the electronic configurations of elements and ions. Look at the examples below for one or two tricky ones.

Worked examples

1 Write out the full electronic configuration of potassium.

$1s^2 2s^2 2p^6 3s^2 3p^6 4s^1$

> This is sometimes written as $[Ar]4s^1$ where $[Ar] = 1s^2 2s^2 2p^6 3s^2 3p^6$ is the configuration of argon.

2 Write out the full electronic configuration of the elements calcium to zinc.

Element	At No.	Electronic configuration
Ca	20	$1s^2 2s^2 2p^6 3s^2 3p^6 4s^2$
Sc	21	$1s^2 2s^2 2p^6 3s^2 3p^6 3d^1 4s^2$
Ti	22	$1s^2 2s^2 2p^6 3s^2 3p^6 3d^2 4s^2$
V	23	$1s^2 2s^2 2p^6 3s^2 3p^6 3d^3 4s^2$
Cr	24	$1s^2 2s^2 2p^6 3s^2 3p^6 3d^5 4s^1$
Mn	25	$1s^2 2s^2 2p^6 3s^2 3p^6 3d^5 4s^2$
Fe	26	$1s^2 2s^2 2p^6 3s^2 3p^6 3d^6 4s^2$
Co	27	$1s^2 2s^2 2p^6 3s^2 3p^6 3d^7 4s^2$
Ni	28	$1s^2 2s^2 2p^6 3s^2 3p^6 3d^8 4s^2$
Cu	29	$1s^2 2s^2 2p^6 3s^2 3p^6 3d^{10} 4s^1$
Zn	30	$1s^2 2s^2 2p^6 3s^2 3p^6 3d^{10} 4s^2$

- Ca: The 4s is lower in energy than the 3d, so the 4s is filled first. The 3d is empty, so is not written down.
- Cr: The electronic configuration of Cr is more stable when the 3d sub-level is half full – $3d^5 4s^1$, than when the 4s is full – $3d^4 4s^2$.
- Cu: The electronic structure of Cu is more stable when the 3d sub-level is completely full – $3d^{10} 4s^1$, than when the 4s is full – $3d^9 4s^2$.

3 Using 'arrows in a box' to represent electrons, write out the electronic configuration of chromium (Cr-24).

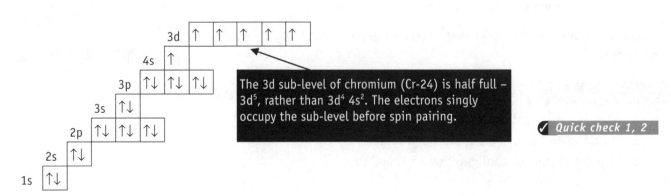

The 3d sub-level of chromium (Cr-24) is half full – $3d^5$, rather than $3d^4 4s^2$. The electrons singly occupy the sub-level before spin pairing.

✓ *Quick check 1, 2*

Electronic configuration of ions

When an atom gains or loses electrons to form an ion, its electronic structure changes. Positive ions are formed by the loss of electrons and negative ions are formed by the addition of electrons.

You can use potassium and chlorine to illustrate what happens.

- The potassium atom loses an electron from its **highest occupied energy level**, (the 4s) forming the K^+ ion.

Potassium atom	→	Potassium ion
K		K^+
$1s^2 2s^2 2p^6 3s^2 3p^6 4s^1$		$1s^2 2s^2 2p^6 3s^2 3p^6$

- The chlorine atom gains one electron, which enters the **lowest occupied** energy level available (the 3p) forming the Cl^- ion.

Chlorine atom	→	Chloride ion
Cl		Cl^-
$1s^2 2s^2 2p^6 3s^2 3p^5$		$1s^2 2s^2 2p^6 3s^2 3p^6$

> Because ions gain and lose electrons the general rule for size is that $r^{2+} < r^+ < r^{atom} < r^- < r^{2-}$ where r = radius.

- In this situation, the K^+ ion and the Cl^- ion have the same electronic structure and are said to be **isoelectronic**.

Transition metal ions are of particular interest.

- The transition metals have a partially filled 3d sub-level, but it is the **4s** electrons that are **lost first** when they form ions.
- The 4s sub-shell still exists, but it is empty, so is not included in the electronic configuration of Cr^{3+}.

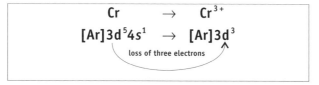

✓ *Quick check 3*

Quick check questions

1. Which sub-level has higher energy in a neutral transition metal atom, the 4s or the 3d?
2. Use the 'arrows in a box method' to illustrate the electronic configuration of **a** copper, **b** silicon.
3. Give the full electronic configurations of **a** Cu^+, **b** Fe^{2+}.

Ionisation energy

First ionisation energy is defined as

> **The amount of energy that is required to remove one mole of electrons from one mole of atoms in the gaseous state, to form one mole of unipositive ions**

$$X(g) \rightarrow X^+(g) + e^-$$

There are three factors that affect the size of the ionisation energy:
- the **distance of the outer electrons from the nucleus** – the nearer the outer electrons are to the nucleus, the harder they are to remove and so the larger the ionisation energy;
- the **nuclear charge** – the more positive the nuclear charge, the harder it is to remove electrons and so the larger the ionisation energy;
- the **electronic shielding** – the lower the amount of shielding between the outer electrons and the nucleus, the stronger the effective nuclear charge will be and the harder it is to remove the outer electrons and so the larger the ionisation energy.

✓ *Quick check 1*

Successive ionisation energies

As more and more electrons are removed from a species, so the amount of energy required to remove the next electron increases. This means that the nth electron is harder to remove than the first because...
- the ion has (increasingly) more protons than electrons, and the same number of protons attract a lower number of electrons;
- the outer electrons are therefore pulled closer to the nucleus – positive ions are smaller than atoms and the more positive the ion becomes, the smaller the ion becomes;
- there is less repulsion between the electrons which remain.

Evidence for electronic structure

The graph shows the energy that is required to remove one electron after another from a mole of potassium atoms in the gaseous phase.

Each part of the graph is explained by the electronic structure of the atoms and a consideration of which electrons are being removed. The large steps (A, B and C) are due to the fact that electrons are being removed from another principal energy level.

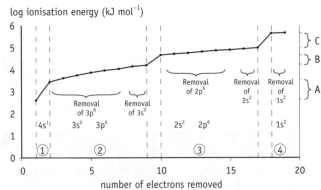

order of removal of electrons
←
$$1s^2\, 2s^2\, 2p^6\, 3s^2\, 3p^6\, 4s^1$$

The electrons fall into four groups. Remember that the higher the ionisation energy, the more difficult the electrons are to remove and so the nearer they are to the nucleus.

1 **The one 4s electron is removed ($4s^1$)**
   This is the easiest electron to remove as it is furthest from the nucleus.

2 **The six 3p electrons are removed ($3p^6$) and then the two 3s electrons are removed ($3s^2$)**
   This is a total of 8 electrons which are in the next shell towards the nucleus.

> The log scale on the y axis just condenses the scale, so we can see the pattern more clearly. Don't worry about it.

These take progressively more energy to remove as the ion has more protons than electrons, so the peripheral electrons are more and more difficult to remove as they are held more strongly. These electrons are shielded less than the $4s^1$ electron.

3 **The six 2p electrons are removed ($2p^6$) and the two 2s electrons are removed ($2s^2$)**
This is a total of 8 electrons which again are in the next shell towards the nucleus, but these are nearer to the nucleus than the 3p and 3s electrons and are harder to remove. Again, these electrons are shielded less than the ones in the 3p and 3s sub-shells, so they experience a larger effective nuclear 'pull' and are therefore more difficult to remove.

4 **The two 1s electrons are removed ($1s^2$)**
These two electrons are the most difficult to remove as they are closest to the nucleus. They experience no 'dilution' or shielding of the nuclear charge by complete inner shells and so require the most energy to remove.

The steps A, B and C (see graph above) are due to the fact that electrons are being removed from different principal energy levels. We therefore have evidence that the electrons surrounding the nucleus are in energy levels and the energy of the principal energy level increases as the distance from the nucleus increases. The diagram represents the increasing energy of the first four principal energy levels of an atom. ▶▶

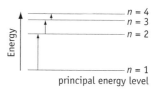

Evidence for sub-shells

The first ionisation energies of the elements of Period 3 provide us with evidence for the existence of sub-shells.

The energy required to ionise the atoms generally increases across the period. This is because…
- the nuclear charge is increasing, but
- electrons are being added to the same shell.

The atoms become smaller because the outer electrons are drawn closer to the nucleus owing to the increase in nuclear charge. The shielding is the same because all elements from sodium to argon have ten electrons in their lower energy levels. As the proton number goes up, the effective nuclear charge increases, making it more difficult to remove the outer electrons.

▶▶ *More about shielding on page 36*

The drop between magnesium and aluminium

The third principal energy level is split into three sub-levels. Here the 3s and the 3p sub-levels contain electrons but the 3p sub-level has a higher energy than the 3s sub-level.
- The electrons in the 3p sub-level are easier to remove than the 3s and will require less energy to remove than the 3s – hence the drop.

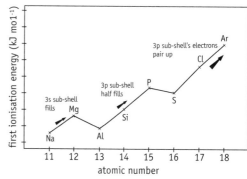

The drop between phosphorus and sulphur

The electrons in the 3p sub-level of phosphorus are all unpaired ($3p^3$), so the extra electron that sulphur has must pair with one of these 3p electrons.
- This electron is easier to remove because of the repulsion of the other electrons in the sub-level.
- This effect is greater than the effect of the increased nuclear charge – hence the drop. ✓ *Quick check 2–4*

Quick check questions

1. Give a full definition of first ionisation energy.
2. How many sub-shells does the third energy level have?
3. Why is aluminium's first ionisation energy lower than magnesium's?
4. Explain why chlorine has a higher first ionisation energy than sulphur.

Module 1: Atomic structure, bonding and periodicity

Using moles

Atoms are too small to weigh out individually, so moles have been invented to make all the numbers involved just that bit easier to understand.

> Learn this set of definitions.

The mole and the Avogadro constant

You will know that a mole is defined as...

> The amount of substance that contains as many particles as exactly 12.000 g of carbon-12

- The number of atoms in 12.000 g of carbon-12 is **6.02×10^{23}**.
- This number is known as the **Avogadro constant**.
- Its units are **particles mol^{-1}**.
- One mole of anything contains **6.02×10^{23}** ...anythings.
- The relative atomic or molecular mass in grams is the mass of one mole of the substance.

Relative atomic mass (RAM or A_r)

The **relative atomic mass** (**RAM** or A_r) of an element is the average mass of the atoms compared to a twelfth of the mass of an atom of carbon-12. This average takes into account the relative abundance of the isotopes that exist.

$$\text{The relative atomic mass} = \frac{\text{the average mass per atom of an atom} \times 12}{\text{the mass of one atom of }^{12}\text{C}}$$

Relative molecular mass (RMM or M_r)

For compounds or molecular ions, the average mass of the molecules or ions is known as the **relative molecular mass** (**RMM** or M_r)

$$\text{The relative molecular mass} = \frac{\text{the average mass per molecule of the entity} \times 12}{\text{the mass of one atom of }^{12}\text{C}}$$

> Entity means, whatever you are talking about e.g. compound or molecular ion.

Working out relative molecular masses

- To work out the M_r of a 'molecule' work out how many atoms of each element are present by looking at the formula, then add up the individual A_r values – simple...

> This even counts for ionic substances, which you and I know aren't really discrete molecules.

Worked examples

Name of substance	Formula	Adding up the A_r values	M_r
Sodium chloride	NaCl	= (1 × Na) + (1 × Cl) = (1 × 23) + (1 × 35.5)	58.5
Sodium carbonate	Na$_2$CO$_3$	= (2 × Na) + (1 × C) + (3 × O) = (2 × 23) + (1 × 12) + (3 × 16)	106
Magnesium hydroxide	Mg(OH)$_2$	= (1 × Mg) + (2 × O) + (2 × H) = (1 × 24) + (2 × 16) + (2 × 1)	58

✓ *Quick check 1*

Converting mass to moles and moles to mass

$$\text{number of moles} = \frac{\text{mass of sample}}{\text{mass of one mole}} = n = \frac{m}{M_r}$$

> Learn this relationship.
>
> ✓ Quick check 2,3,4

Worked example

Find the number of moles of calcium carbonate ($CaCO_3$) in a 7.5 g sample.

Step 1 Write down what you know.
number of moles = mass ÷ relative molecular mass ∴ $n = m/M_r$
n = ?
M_r = Work out
m = 7.5 g

Step 2 Work out the other unknown *(i.e. the relative mass of the sample M_r)*.
M_r = (1 × Ca) + (1 × C) + (3 × O)
    = (1 × 40) + (1 × 12) + (3 × 16) = 100

Step 3 Calculate the number of moles, n
$n = m/M_r = 7.5/100 = 0.075$ moles
∴ 7.5 g of $CaCO_3$ contain <u>0.075 moles</u>

Equations

- **Equations** just tell us what substances are reacting and what products are made. A **balanced equation** tells us in what **proportions** substances react and are made. There are three types of equations you need to know about.
- **Molecular equations** show the proportions and the formulae of the reactants and products.
- **Ionic equations** show the main ions present in the reaction.
- **Half equations** (or **ion–electron equations**) show the reduction or oxidation processes in a reaction.

> You must be able to balance all types of equations as matter can't be created or destroyed.
>
> In half equations **the overall charge** on each side of the equation must be the same.

❓ Quick check questions

1. Work out the RMM of **a** $ZnCl_2$, **b** NaOH, **c** H_2O_2.
2. Work out the mass of 1 mole of **a** calcium hydroxide, **b** copper(II) chloride, **c** ammonium nitrate.
3. What is the mass of **a** 0.1 moles of sulphuric acid, **b** 0.25 moles Na_2SO_4.
4. How many moles are there in 10 g of **a** $CaCO_3$, **b** NaOH, **c** $KMnO_4$.
5. Write out a balanced molecular equation for the complete combustion of pentane (C_5H_{12}).

Module 1: Atomic structure, bonding and periodicity

Empirical formula

Empirical formula is defined as...

> **The simplest whole number ratio of atoms of each element in a compound**

It does not represent the actual number of atoms in a compound, only the **ratio** of the elements that are present. For example, propene has the molecular formula, C_3H_6, but the empirical formula is CH_2.

Molecular formula is defined as...

> **The number of atoms of each element in one molecule of a compound**

For example, butane has the molecular formula C_4H_{10} meaning each molecule has 4 carbons and 10 hydrogens present. (Its empirical formula would be C_2H_5.)

The molecular formula is a multiple of the empirical formula

> **empirical formula × n = molecular formula**

E.g. $(C_2H_5) \times 2 = C_4H_{10}$

and

> **empirical formula mass × n = molecular formula mass**

E.g. $((2 \times C) + (5 \times H)) \times 2 = C_4H_{10}$

$((2 \times 12) + (5 \times 1)) \times 2 = C_4H_{10}$

$29 \times 2 = 58$

> n is a whole number, an integer.

✓ *Quick check 1, 2*

Empirical formula is usually calculated from experimental data

The most common questions come from percentage composition data which is given to you in the question – the data come from experimental work. You need to follow the general method below and then apply it to the worked examples and questions. It may need fine-tuning depending on the question asked.

> Step 1 Write down the information you are given
> Step 2 Convert the masses to moles using, $n = m \div A_r$ (or $n = m \div M_r$)
> Step 3 Express these answers as the molar ratio
> Step 4 Simplify the ratio – by dividing by the smallest number from step 3 to give the empirical formula
> Step 5 Smile ☺

> number of moles = $\dfrac{\text{mass of sample}}{\text{mass of one mole}}$,
> so $n = m \div A_r$

Look at the example to see how this method works.

Amount of substance

Worked examples

1 Calculate the empirical formula of the compound formed when 3.84 g of magnesium react with 11.36 g of chlorine.

	magnesium	chlorine
Step 1 Write down the information you have been given (*i.e. the mass of each element*)	3.84 g	11.36 g
Step 2 Convert the mass to moles	$n = m \div A_r$ $\frac{3.84}{24} = 0.16$	$n = m \div A_r$ $\frac{11.36}{35.5} = 0.32$
Step 3 Express these answers as a molar ratio	0.16	0.32
Step 4 Simplify the ratio – dividing by the smallest number from step 3 to give the empirical formula	$\frac{0.16}{0.16} = 1$	$\frac{0.32}{0.16} = 2$
	Mg	Cl_2
Step 5	☺ Ha!	

2 Analysis of a chloride of sulphur, Z, was found to have 47.4% sulphur by mass. In the mass spectrum of Z the largest value of *m/z* occurs at 135. Calculate the empirical and molecular formulae.

	sulphur	chlorine
Step 1 Write down the information you have been given (*i.e. the % of each element*)	47.4%	100 − 47.4 = 52.6%
Step 2 Convert the percentage to moles	$n = m \div A_{rI}$ $\frac{47.4}{32} = 1.48$	$n = m \div A_r$ $\frac{52.6}{35.5} = 1.48$
Step 3 Express these answers as a molar ratio	1.48	1.48
Step 4 Simplify the ratio – dividing by the smallest number from step 3 to give the empirical formula	$\frac{1.48}{1.48} = 1$	$\frac{1.48}{1.48} = 1$
	S_1Cl_1 or	SCl
Step 5 Work out the molecular formula	empirical formula mass × n = molecular formula mass $(32 + 35.5) \times n = 135.0$ $n = 135.0/67.5 = 2$ molecular formula of **Z** is S_2Cl_2	
Step 6	☺ Double Ha!	

▸ Assume that the sample weighs 100g

▸ Adapt the method, as you are asked to find the molecular formula as well

✓ *Quick check 3, 4*

Quick check questions

1. Define empirical formula and molecular formula.
2. What is the relationship between empirical formula and molecular formula?
3. A compound was found to have 52.1% carbon, 13.0% hydrogen and 34.8% oxygen. Calculate its empirical formula.
4. A dodgy liquid from an old refrigerator found on a scrap heap was found to contain a compound with the following composition: 11.4% carbon, 34.0% chlorine and 54.6% fluorine. Calculate its empirical formula.

Reacting amounts

You may be asked to calculate how much product you get from a certain amount of reactant, or how much reactant is needed to make a certain amount of product. The second of these is very important in industry when calculating how much product could be made from a certain amount of feedstock.

Make sure you know this equation below, as it is the cornerstone of many calculations.

$$\text{Number of moles} = \frac{\text{mass of sample}}{\text{mass of one mole}}$$

$$n = m/M_r$$

▶ n = mol
m = g
M_r = g mol^{-1}

✓ *Quick check 1*

Worked example

1 What mass of iron could be obtained from 320 tonnes of iron(III) oxide by reaction with carbon monoxide in the blast furnace?

The method for working out this calculation out involves...
1. *working out the number of moles of reactant from a given mass of reactant;*
2. *looking at the stoichiometry of the equation to get the molar ratio;*
3. *working out the mass of product from the number of moles of product.*

Steps 1–3 can be summarised in the flow diagram below

❷ equation
↑   ↓
❶ mass→moles   ❸ moles→mass
 for a reactant   *for a product*

▶ 1 tonne = 1000 kg

Step 1 Write down what you know and calculate the number of moles of iron(III) oxide present.

$$Fe_2O_3(l) + 3CO(g) \rightarrow 2Fe(l) + 3CO_2(g)$$
320 tonnes   ?

320 tonnes = 320,000 kg = 320,000 × 1000 g = 320,000,000 g
Number of moles?

$$n = m/M_r \qquad n = ?$$
m = 320,000,000 g
M_r = (2 × Fe) + (3 × O) = (2 × 56) + (3 × 16) = 160
n = 320,000,000/160 = **2,000,000 moles of Fe$_2$O$_3$**

Step 2 Look at the equation and work out the number of moles of iron produced.

1Fe$_2$O$_3$ ≡ 2Fe
1 mole → 2 moles

∴ If we have 2,000,000 moles of Fe$_2$O$_3$ we will get 2 × 2,000,000 moles of Fe
∴ We get 4,000,000 moles of iron produced

Step 3 Calculate the mass of this 4,000,000 moles of iron.

$$n = m/M_r \therefore m = n \times M_r$$
m = 4,000,000 × 56
m = 224,000,000 g of iron
m = **224 tonnes of iron**

✓ *Quick check 2*

2 Quicklime (calcium oxide) is produced by roasting limestone (calcium carbonate) in a rotating limekiln. The equation for the reaction is:

$$CaCO_3(s) \rightarrow CaO(s) + CO_2(g)$$

Calculate the mass of limestone required to produce 16.8 tonnes of quicklime.

This requires you to perform a calculation using the previous method in reverse. Steps 1–3 in the worked example can be summarised in this flow diagram below

```
              ❷ equation
              ↓         ↑
      ❸ mass←moles  ❶ moles←mass
        for a reactant   for a product
```

Step 1 Write down what you know and calculate the number of moles of calcium oxide.

$$CaCO_3(s) \rightarrow CaO(s) + CO_2(g)$$
$$\quad ? \qquad\qquad 16.8 \text{ tonnes}$$
$$\qquad\qquad\qquad 16{,}800 \text{ kg}$$
$$\qquad\qquad\qquad 16{,}800 \times 1000 \text{ g}$$

Moles of $CaCO_3$?

$$n = m/M_r \qquad n = ?$$
$$m = 16{,}800{,}000 \text{ g}$$
$$M_r = (1 \times Ca) + (1 \times O) = (1 \times 40) + (1 \times 16) = 56$$
$$n = 16{,}800{,}000/56 = 300{,}000 \text{ moles of CaO present}$$

Step 2 Look at the equation and work out the number of moles of iron produced.

$$CaO \equiv CaCO_3$$
$$1 \text{ mole} \rightarrow 1 \text{ mole}$$

∴ If we have 300,000 moles of CaO we will need 300,000 moles of $CaCO_3$.

Step 3 Calculate the mass of 300,000 moles of calcium oxide.

$$n = m/M_r \therefore m = n \times M_r$$
$$M_r = (1 \times Ca) + (1 \times C) + (3 \times O)$$
$$\quad = (1 \times 40) + (1 \times 12) + (3 \times 16) = 100$$
∴
$$m = n \times M_r$$
$$m = 300{,}000 \times 100$$
$$m = 30{,}000{,}000 \text{ g of } CaCO_3 = \textbf{30 tonnes of } CaCO_3$$

So <u>30 tonnes</u> of limestone are required to produce <u>16.8 tonnes</u> of quicklime.

✓ *Quick check 3*

Quick check questions

1 Which of the following contains the largest number of moles: **a** 10 g of NaOH, **b** 15 g of H_2SO_4, **c** 20 g of $CaCO_3$?

2 What is the maximum mass of sodium chloride which could be obtained by reacting 5.3 g of dry sodium carbonate in excess acid as shown below?

$$Na_2CO_3(s) + 2HCl(aq) \rightarrow 2NaCl(aq) + CO_2(g) + H_2O(l)$$

3 Titanium is produced industrially by reaction of titanium(IV) chloride with sodium between 500 and 1000°C in an inert atmosphere.

$$TiCl_4 + 4Na \rightarrow Ti + 4NaCl$$

Calculate the mass of titanium(IV) chloride required to produce 14.4 g of titanium. (A_r (Ti) = 48)

Reacting volume calculations

In these calculations you calculate the volume of gas used or produced in a reaction.

The molar gas volume

At the same temperature and pressure equal volumes of gases contain equal numbers of molecules (or atoms). This is **Avogadro's law.**

> One mole of any gas occupies 22.4 dm^3 at standard temperature and pressure (stp)

> number of moles = $\dfrac{\text{number of dm}^3 \text{ of the gas (at stp)}}{22.4}$ = n = vol/22.4

> stp is...standard temperature and pressure:
> temperature = 273 K
> pressure = 100 kPa

✓ Quick check 1, 2

Worked example

Pentane burns in oxygen as shown in the equation below.

$$C_5H_{12}(g) + 8O_2(g) \rightarrow 5CO_2(g) + 6H_2O(g)$$

What is the minimum volume of oxygen gas needed for the complete combustion of 20 dm^3 of propane at stp?

Steps 1-3 of the worked example can be summarised in this flow diagram

```
              ❷ equation
              ↑      ↓
  ❶ volume→moles  ❸ moles→volume
     reactant        reactant
      C₅H₁₂            O₂
```

Step 1 Write down what you know and calculate the number of moles of pentane present.

$$C_5H_{12}(g) + 8O_2(g) \rightarrow 5CO_2(g) + 6H_2O(g)$$
$$20 \text{ dm}^3 \quad\quad ?$$

No. of moles present?

$$n = \text{vol}/22.4$$
$$20/22.4 = 0.89 \text{ moles}$$

Step 2 Look at the equation and work out the number of moles of oxygen needed.

$$1C_5H_{12} \equiv 8O_2$$
$$1 \text{ mole} \rightarrow 8 \text{ moles}$$

∴ If we have 0.89 moles of C$_5$H$_{12}$ we will need 8 × 0.89 moles of O$_2$.
∴ We need 7.12 moles of oxygen.

Step 3 Calculate the volume of 7.12 moles of oxygen.

$$n = \text{vol}/22.4 \quad \therefore \text{vol} = n \times 22.4$$
$$\therefore \text{vol} = 7.12 \times 22.4 = \underline{160 \text{ dm}^3}$$

To calculate a volume of a gas from a given mass of a gas just use this nifty equation.

> $\dfrac{\text{volume of gas (in dm}^3)}{22.4} = \dfrac{\text{mass of gas}}{M_r \text{ of gas}}$

> Remember: always use SI units.
> • 1 m^3 = 1 × 10^6 cm^3 = 1 × 10^3 dm^3 (so cm^3 = 10^{-6} m^3 = 10^{-3} dm^3)
> • 1 atmosphere = 10^5 Pa = 100 kPa
> • 760 mmHg = 100 kPa
> • 273 K = 0°C (°C are not used in calculations)

Worked example

What volume would 10 g of hydrogen have at stp?

$$\frac{\text{Volume of gas (in dm}^3)}{22.4} = \frac{\text{Mass of gas}}{M_r \text{ of gas}}$$

Step 1 Put values given into the equation. $\frac{Vol}{22.4} = \frac{10}{2}$

Step 2 Rearrange to find the unknown. $Vol = 22.4 \times \frac{10}{2} = \underline{112 \text{ dm}^3}$

> *Assumptions of the Kinetic Theory of Gases*
> Particles...
> - are hard spheres;
> - have negligible size;
> - move in rapid and random motion;
> - have no intermolecular forces.

The ideal gas equation

Ideal gases obey the assumptions of the **Kinetic Theory of Gases**.

The ideal gas equation came about by considering a number of relationships:

- **Boyle** found that at a constant temperature, the volume of a gas is inversely proportional to its pressure.

$$V \propto 1/p$$

- **Charles** found that at constant pressure, the volume of a gas is directly proportional to its absolute temperature.

$$V \propto T$$

- **Avogadro's Law** can be rearranged to state that at a constant temperature and pressure, the volume of any gas is directly proportional to the number of moles of the gas.

$$V \propto n$$

Combining these we get...

$$V \propto \frac{nT}{p} \text{ or } V = \frac{RnT}{p}$$

where R is a proportionality constant.

> p = pressure in Pa (pascals)
> V = volume in m³ (metres cubed)
> R = the gas constant, 8.31 J K⁻¹ mol⁻¹ (joules per kelvin per mole)
> T = temperature measured in K (kelvin)
> n = number of moles and remember it's Avogadro not Avocado...

or

The ideal gas equation $pV = nRT$

✓ *Quick check 3*

Worked example

Calculate the volume which 2.2 g of carbon monoxide would occupy at a temperature of 20°C and a pressure of 6500 Pa.

Step 1 Write down what you know.
$PV = nRT$

Step 2 Put the figures into the equation.
$6500 \times V = 2.2/28 \times 8.31 \times 293$

Step 3 Rearrange to solve the unknown.
$$V = \frac{2.2/28 \times 8.31 \times 293}{6500}$$
$V = 0.02943 \text{ m}^3 = \underline{29.43 \text{ dm}^3}$

Step 4 Double check your units.
All SI units ✓

$p = 6500$ Pa
$V = ?$ m³
$n = m/M_r = 2.2/28$
$R = 8.31$ J K⁻¹ mol⁻¹
$T = 20°C = 20 + 273 = 293$ K

> Always make sure the units used are the correct ones.

> Remember that density is m/V. This is sometimes substituted into the ideal gas equation to give $pM_r/RT = m/V$

✓ *Quick check 4*

❓ Quick check questions

1. What is stp?
2. What is the volume of 1 mole of any gas at stp?
3. What is the volume of 0.75 moles of SO_2 at stp?
4. What is the volume of 2.50 g of hydrogen gas at a temperature of 293 K and a pressure of 100 kPa?

Module 1: Atomic structure, bonding and periodicity

Molarity and concentration

Molarity and **concentration** tell us about the amount of a substance in a particular volume of another substance. It's measured in moles per dm^3. For example, a 1 molar solution of sodium chloride has one mole of NaCl in 1 dm^3 of solution.

1 mole of NaCl is $(1 \times Na) + (1 \times Cl) = (1 \times 23) + (1 \times 35.5) = 58.5$ g

So if 58.5 g of NaCl was put into enough water to make 1 dm^3 of solution, you would have a 1 molar solution – 1 M = 1 mole per dm^3 or 1 mol dm^{-3}
For a given solution...

> **number of moles = concentration × volume**
> $n = c \times V$

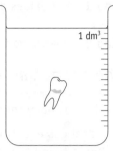

A Molar solution?

where n = mol, c = mol dm^{-3}, V = dm^3.
Some books and teachers prefer to use 'molarity' instead of 'concentration'.
∴ **number of moles = molarity × volume**
$n = M \times V$

▷ concentration = No. of moles/volume (units = mol dm^{-3})

> **number of moles** = $\dfrac{\text{concentration} \times \text{volume}}{1000}$
> $n = c \times V/1000$

where n = mol, c = mol dm^{-3}, V = cm^3.

∴ **number of moles** = $\dfrac{\text{molarity} \times \text{volume}}{1000}$

$n = M \times V/1000$

▷ We divide by 1000 as the volume is in cm^3.

Worked example

Calculate the molarity of the solution made by adding 5 g of solid NaOH to enough water to make 25 cm^3 of solution.

Step 1 Calculate the number of moles of sodium hydroxide present.
$n = m/M_r$ M_r (NaOH) = $(1 \times Na) + (1 \times O) + (1 \times H)$
$= (1 \times 23) + (1 \times 16) + (1 \times 1)$
$= 40$
∴ $n = m/M_r$
$= 5/40 = 0.125$ moles

▷ Molarity is easy when you get your teeth into it – try chewing your way through the worked examples on these pages.

Step 2 Calculate the concentration of the solution.
number of moles = $\dfrac{\text{concentration} \times \text{volume}}{1000}$

$n = \dfrac{c \times V}{1000}$

$0.125 = \dfrac{c \times 25}{1000}$

$c = \dfrac{0.125 \times 1000}{25}$

$= 5$ mol dm^{-3}

The solution is a 5 M solution

✓ *Quick check 1*

Titration calculations

Titrations involve either acid/base or redox reactions. Questions usually ask you to calculate one of the following...

- the concentration of a reactant;
- the concentration of a product;
- the volume of a reactant;
- the volume of a reactant;
- to balance an equation.

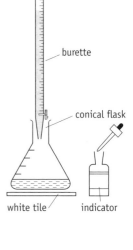

Worked example

25.0 cm³ of a 0.5 M solution of sodium hydroxide was titrated against a solution of nitric acid. 15.0 cm³ of the acid was required to neutralise the alkali. Calculate the concentration of the acid and the concentration of sodium nitrate produced.

Step 1 Write down what you know and calculate the number of moles of sodium hydroxide.

$$NaOH(aq) + HNO_3(aq) \rightarrow NaNO_3(aq) + H_2O(aq)$$
$c = 0.5$ M $c = ?$ $c = ?$
$V = 25$ cm³ $V = 15$ cm³

Moles of NaOH

$$n = \frac{cV}{1000} = \frac{0.5 \times 25}{1000} = 0.00125 \text{ moles}$$

Step 2 Look at the equation to find the molar ratio.

$$1NaOH \equiv 1HNO_3$$
$$1 \text{ mole} \rightarrow 1 \text{ mole}$$

∴ We have 0.00125 moles of NaOH so we will need 0.00125 moles of HNO_3 to neutralise it.

Step 3 Calculate the concentration of the acid.

$$n = \frac{cV}{1000}$$
$$0.00125 = \frac{c \times 15}{1000}$$

Rearrange

$$c = \frac{0.00125 \times 1000}{15} = \underline{0.08 \text{ mol dm}^{-3}}$$

> You should only be asked questions about monoprotic acids like HCl or HNO_3 – if they stick H_3PO_4 in a question they're being a bit tough.

> Steps 1–3 of the worked example can be summarised in this flow diagram:
>
> ❷ equation
> ↑   ↓
> ❶ c and $V \rightarrow$ moles   ❸ moles and $V \rightarrow$ ❹ c
> for reactant NaOH   for reactant HNO_3 and product $NaNO_3$

Step 4 Calculate the concentration of $NaNO_3$.

We have 0.00125 moles of NaOH so the equation tells us that we will have 0.00125 moles $NaNO_3$ in 25 + 15 = 40 cm³ *(ignore the volume of water produced in the reaction)*.

concentration = moles/volume
We have 0.00125 in 40 cm³ of water

∴ in 1 dm³ of water there would be $1000/40 \times 0.00125 = \underline{0.0313 \text{ mol dm}^{-3}}$.

Quick check questions

1. Calculate the molarity of hydrochloric acid in each of these cases:

   a 1 mol made up to 100 cm³ of solution, **b** 0.05 mol made up to 25 cm³ of solution.

2. Calcium carbonate reacts with hydrochloric acid as shown below:

   $$CaCO_3(s) + 2HCl(aq) \rightarrow CaCl_2(aq) + H_2O(l) + CO_2(g)$$

   Calculate the volume of 2 M HCl needed to react completely with 5 g of $CaCO_3$.

Bondage

They say that opposites attract, and in the case of ionic bonding they're about right. I'm sure you remember that metals form positive ions and that non-metals form negative ions. Unsurprisingly, these two types of ions attract each other to make **ionic substances** – some of which are great on fish 'n' chips.

Ionic bonding

Ionic bonds are formed by the **electrostatic forces of attraction** between two oppositely charged ions.

- Ions are formed by **electron transfer**.
- The metal loses one or more electrons and the non-metal gains one or more electrons.
- Most atoms try to obtain a **full outer shell** as it is very stable.
- Ionic bonds are **non-directional**.

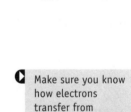

Dot and cross diagrams show what's happening

Example: Sodium chloride – NaCl

- Sodium chloride is composed of Na^+ and Cl^- ions.
- The Na^+ ion is formed from a Na atom when the atom loses an electron.

$$Na \rightarrow Na^+ + e^-$$

- The Cl^- ion is formed from a Cl atom when the atom gains an electron.

$$Cl + e^- \rightarrow Cl^-$$

- The Na atom has one electron in the outermost 3s sub-level. It loses this to the Cl atom which has seven electrons in its outermost 3s and 3p shells.
- The result is that each atom attains **full shell stability** by becoming an **ion**.

$$\begin{array}{lll} \text{atoms} & \rightarrow & \text{ions} \\ Na \;\; 1s^2 2s^2 2p^6 3s^1 - e^- \rightarrow & Na^+ \;\; 1s^2 2s^2 2p^6 \\ Cl \;\; 1s^2 2s^2 2p^6 3s^2 3p^5 & \rightarrow & Cl^- \;\; 1s^2 2s^2 2p^6 3s^2 3p^6 \end{array}$$

> Make sure you know how electrons transfer from magnesium to oxygen as well.

✓ *Quick check 1*

Properties of ionically bonded substances

Ionic compounds have a giant lattice structure, in which each ion is surrounded by ions of opposite charge. The ions are held in place by the electrostatic forces of attraction that exist between them. Ionic bonds are strong, which means that ionic substances:

- have high melting points;
- have high boiling points;
- are soluble in water;
- conduct electricity when molten or dissolved in water.

> Remember that's giant lattice, not giant lettuce...

Covalent bonding

Covalent bonds generally form when non-metal elements bond. The bond involves the atoms **sharing pairs of electrons**.
- Usually an electron from each atom form the shared pair of electrons.
- The bond forms because each atom effectively gains another electron.
- Therefore atoms often attain a full shell.
- The bond strength can vary from ~150 to 900 kJ mol^{-1}.
- We can show covalent bonding by 'dot-and-cross' diagrams or lines between the symbols of two elements.

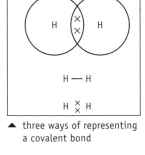

▲ three ways of representing a covalent bond

Dot and cross diagrams

Example: Chlorine gas, Cl_2.
- Both Cl atoms have 7 electrons in the outer (3s and 3p) sub-shells.
- These are arranged in 3 pairs + 1 single electron.
- Two Cl atoms combine so that each shares the single electron of the other.
- Both chlorine atoms therefore have a full principal energy level.

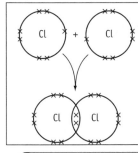

✓ *Quick check 2*

Multiple bonds

It is possible for two atoms to share more than one pair of electrons.
- Sharing two pairs of electrons gives a **double** covalent bond.
- Sharing three pairs of electrons gives a triple bond. (Don't worry about these.)

Dative covalent bonding

A dative covalent bond is a special type of covalent bond where both the electrons in the bond are supplied by **one atom** only (example 1 below).
- Dative covalent bonds are sometimes known as **co-ordinate** bonds.
- Dative covalent bonding can occur between molecules (example 2 below).
- Pairs of molecules are held together to form **dimers**.
- The process of making a dimer is cunningly called **dimerisation**.

Examples

1 NH_4^+ has dative covalent bonding in one of the bonds between the nitrogen and the hydrogen.

2 Solid aluminium chloride has the formula Al_2Cl_6. This happens when vaporised $AlCl_3$ molecules are cooled. As Al_2Cl_6 is **two** identical molecules joined up, it is called a dimer.

▶▶ *Metallic bonding is covered on page 30*

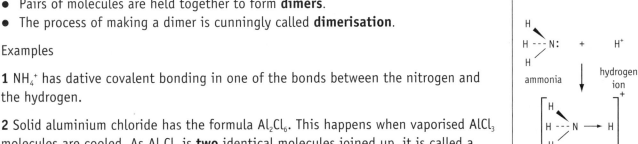

✓ *Quick check 3*

Quick check questions

1 Draw dot-and-cross diagrams to show the bonding or the transfer of electrons in:
   a NaBr, **b** H_2O, **c** $MgCl_2$, **d** H_3O^+, **e** O_2, **f** N_2.
2 State whether the bonding is ionic or covalent in each case **1a** → **1f**.
3 For the covalent compounds in **1a** → **1f**, draw the molecule with a single line representing one covalent bond.

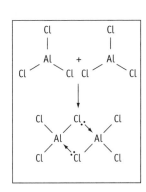

Bond polarity

In a chlorine molecule the atoms bonded together are the same. The electrons are shared equally between the two, as the atoms attract the electrons to the same extent. However, when elements with different **electronegativities** bond together, the electrons are unequally shared. This gives rise to **polar bonds**.

> Electronegativity is defined as the ability of an atom to attract a pair of electrons in a covalent bond. δ = delta, which means 'a small amount'

Causes of bond polarity

Unequal sharing of bonding pairs of electrons causes a covalent bond to be **polar**.
- Polar means that there is a **charge separation** in the bond.
- One atom will be slightly positive ($\delta+$) and **electron deficient** and the other will be slightly negative ($\delta-$) and **electron rich**.
- The most **electronegative** atom attracts the bonding electrons the most.
- A polar molecule acts a bit like a **small bar magnet**, except that the attraction is due to electric charges, not magnetic poles.
- The molecule is said to have a **permanent dipole** and the bond is said to have a **dipole moment**.
- This symbol is often used to signify a **permanent dipole** $\longmapsto$.

✓ Quick check 1

Factors affecting electronegativity

There are two factors that determine the electronegativity of an element: the **nuclear charge** and the **atomic radius**.

1 The larger the nuclear charge
   - the more it will attract bonding pairs of electrons;
   - the larger will be the electronegativity.

2 The smaller the atomic radius
   - the greater will be the attraction between the nucleus and the bond electrons;
   - the less will be the shielding effect;
   - the larger will be the electronegativity.

> Think of atomic radius as half the distance between the centres of two adjacent atoms. (See page 35 for more details.)

On descending a group, the effective nuclear charge on an atom remains the same but the size increases. Therefore the top member is more electronegative, as it is smaller and attracts the bonding electrons more.

Decrease in atomic radius and increase in nuclear charge leads to an increase in electronegativity.

H 2.1							He –
Li 1.0	Be 1.5	B 2.0	C 2.5	N 3.0	O 3.5	F 4.0	Ne –
Na 0.9	Mg 1.2	Al 1.5	Si 1.8	P 2.1	S 2.5	Cl 3.0	Ar –
K 0.8						Br 2.8	
Rb 0.8						I 2.5	
Cs 0.7						At 2.2	

Decrease in atomic radius and degree of shielding leads to an increase in electronegativity.

✓ Quick check 2

Bonding

Polarisation of ions

As you know, ionic substances are formed from ions of metals and non-metals. If the metallic cations (+ve ions) lie very close to the non-metal anions (-ve), the positive cations can attract the electron cloud on the anion and distort it. The negative ion is then said to be **polarised** by the cations. This results in the ionic compound being **slightly covalent** in nature, as there is some sharing of electrons.

Electrons on the negative ion are drawn towards the positive ion if...

- the positive ion is small
- the positive ion is highly charged } high **charge density**
- the negative ion is large

Charge density describes the spread charge over the ion.

Good examples of cations with high charge densities are found at the top of groups I and II and III like Li^+, Be^{2+}, Mg^{2+} and Al^{3+}.

Facts you should know about bonding

- If the electronegativity difference between the elements is small, the bonding is covalent (a difference of 0 = perfectly covalent).
- If the electronegativity difference between the elements is large, the bonding is ionic (a difference >2.1 = ionic).
- If the positive ion in an ionic bond is very polarising, the bonding is ionic with some covalent nature (difference between 0 and 2.1).

To summarise: Bonding is not as simple as just ionic on one side and covalent on the other. There is a kind of gradation between the two.

	ionic	polarised ions	polar covalent	covalent
Example	NaCl	$MgCl_2$	$AlCl_3$	$SiCl_4$
Electronegativity difference	2.1	1.8	1.5	1.2

Quick check questions

1. Define the term electronegativity.
2. State and explain how electronegativity changes across a period and down a group.
3. Explain how negative ions can become polarised.
4. Draw the charge separation on the following molecules **a** HCl, **b** HF, **c** H_2O.
5. Why is silicon tetrachloride covalent but sodium chloride is ionic?

Intermolecular forces

At GCSE level you learnt about intramolecular forces of attraction or bonds which occur 'inside' or 'within' molecules or ionic substances. These hold the particles together to form molecules. But there are other attractive forces called **intermolecular forces**. These arise 'between' molecules and can be thought of as holding individual molecules together. These weaker forces must be overcome for a substance to boil. There are three types you must know about.

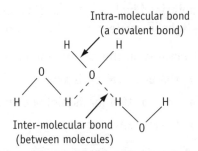

1 van der Waals forces

van der Waals forces (VDWs) are surface forces and arise from the fluctuating movement of electrons. This movement causes temporary dipoles, which, in turn, induce attraction between molecules – it's easy, read on.
- **Temporary dipoles** occur in one molecule owing to an unequal distribution of the electron cloud around it, which may happen from time to time.
- This temporary dipole induces a dipole in another molecule and the two attract each other.
- van der Waals forces are also known as **induced dipole–dipole** forces.
- Their strength is between 1 and 20 kJ mol^{-1}.

The strength of van der Waals forces increases
- the larger the molecule (due to a larger electron cloud);
- the larger the surface area of the molecule (due to a larger electron cloud exposed).

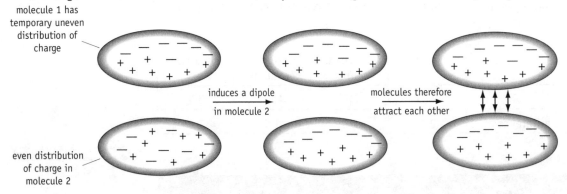

van der Waals forces affect physical properties such as...
- boiling point
- melting point
- viscosity.

The more van der Waals forces exist between molecules, the more 'sticky' they become as they attract each other more. Energy is required to overcome these forces of attraction.

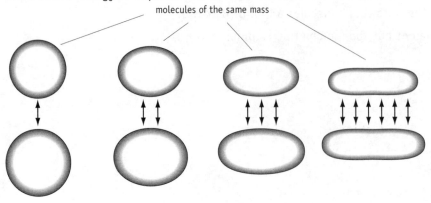

2 Permanent dipole–dipole forces

These forces occur between molecules with a **permanent dipole**. This happens when the electronegativities of the elements are very different, resulting in a polar bond.
- A δ+ atom of one molecule attracts the δ– atom of another molecule.
- This results in a weak electrostatic force of attraction.
- Permanent dipole–dipole forces are usually stronger than van der Waals forces.
- The strength is 5–20 kJ mol^{-1}.

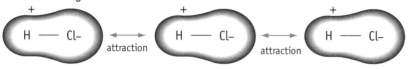

permanent dipoles cause intermolecular attraction

equal pull on electron evens out, so no dipole on either molecule

- No overall dipoles are seen in symmetrical molecules like carbon dioxide owing to the 'equal pull' on the bonding electrons.

> Oxygen and chlorine both have an electronegativity of 3.0, but HCl does **not** contain hydrogen bonding as the chlorine is too large to approach the hydrogen (but it does have a permanent dipole).

3 Hydrogen bonding

This is the strongest intermolecular force and is a special case of a permanent dipole–dipole interaction.
- It *only* occurs where hydrogen is bonded to **nitrogen, oxygen** or **fluorine**.
- Nitrogen, oxygen and fluorine are very electronegative and cause the bond with hydrogen to be polar.
- The bond is so polarised that the hydrogen forms a weak bond with the nitrogen, oxygen or fluorine of another molecule.
- Nitrogen, oxygen and fluorine are small enough to approach a hydrogen atom – the lone pair on the N, O or F forms a bond with the δ+ hydrogen.
- Hydrogen bonds are represented by three dashes (– – –).
- The strength is 20–40 kJ mol^{-1}.

the lone pair on the oxygen is attracted to the very small hydrogen

the hydrogen in water has little share of the e$^-$ in the covalent bond, so is δ+

hydrogen bond ammonia

Examples of molecules which are hydrogen bonded are shown above.

✓ *Quick check 1*

✓ *Quick check 2, 3*

Quick check questions

1. Name the three types of intermolecular force and describe how they occur.
2. Why does pentane (C_5H_{12}) have a boiling point of 36.3°C but 2,2-dimethylpropane ((CH_3)$_4$C) has a boiling point of 9.5°C?
3. Why does water have a boiling point of 100°C but hydrogen sulphide, which has heavier molecules, has a boiling point of –61°C?

Module 1: Atomic structure, bonding and periodicity

States of matter

What's small and hard to glue together: the particle model... yes it's solids, liquids and gases yet again. Make sure that you know: (1) what the particle model is, (2) how the energy of particles changes on heating and (3) how the distance between particles changes on heating.

> Some books call a 'state' a 'phase', so a solid changes 'phase' to a liquid.

The particle model

The **particle model**, or **kinetic theory of matter**, considers matter to be solid, liquid or gas and to be made of particles. The particles can be atoms, ions or molecules.

Solids
- have fixed shapes;
- are highly **ordered**;
- have particles close together;
- are **densely** packed;
- are held together by inter- and intramolecular forces of attraction;
- have particles which vibrate about a fixed position, but are not free to move and therefore have no **translational energy**.

a simple model of solid

Liquids
- take the shape of the container that they are in;
- have some degree of order but less than solids;
- have particles close together;
- have densities slightly less than solids (except water/ice);
- have particles which are more free to move than in solids. They are still held together by similar forces, but the forces have been weakened.

a simple model of liquid

Gases
- don't have a fixed shape – they fill the volume available;
- have particles widely spaced;
- have low densities;
- have particles moving in rapid random motion;
- will diffuse by particles moving from high concentration to low concentration.

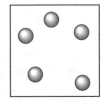

a simple model of a gas

✓ *Quick check 1, 2*

Changing state

1. Melting – a solid to a liquid
   - When a solid is heated the particles gain **energy**.
   - When sufficient energy is gained, the **bonds weaken** or **loosen**.
   - Particles therefore become more free to move around.
   - The energy required to melt a solid is known as the **enthalpy of fusion**.

2. Boiling – a liquid to a gas
   - When a liquid is heated the particles gain energy.
   - Eventually some gain sufficient energy to overcome the **inter-particle forces** of attraction which exist, and some eventually escape to form a gas.

> Sublimation is when a substance changes from a solid straight to a gas or vice versa.

- The energy required to boil a liquid is known as the **enthalpy of vaporisation**. This is normally larger than the enthalpy of fusion, as bonds have to be completely broken.

3 Condensing – a gas to a liquid
- When a gas is cooled the particles lose energy and slow down.
- Eventually the particles start bonding with each other.
- This eventually causes a liquid to form.

4 Freezing (or solidification) – a liquid to a solid
- When a liquid is cooled the particles lose energy and slow down.
- Eventually the particles slow down enough to become **ordered**.
- This is when a solid forms.

Evaporation – liquid to a gas

For pure substances, the physical processes listed 1–4 above occur at a specific temperature, namely at the melting point or boiling point of the substance. This is useful as the information can be used to identify a pure unknown substance.

Evaporation is where a liquid turns to a gas at **any temperature**.

- The most **energetic** particles escape the forces of attraction between the particles and form a gas.
- The **average energy** of the particles remaining is lowered since the most energetic ones leave.

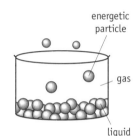

a liquid evaporating

> Physicists may disagree with this definition... but that's physics for you.

Cooling curves

These curves are like the one shown here. You need to understand what's happening in each segment of the graph, especially the flat bits or, as examiners like to call them, **plateau regions**.

What's going on?
1 The gas cools and particles begin to slow down.
2 The gas starts condensing as particles lose energy and get closer together.
3 The liquid cools and particles lose energy and slow down.
4 The liquid freezes or solidifies and particles lose energy and get closer.
5 The solid cools down to the temperature of the surroundings.

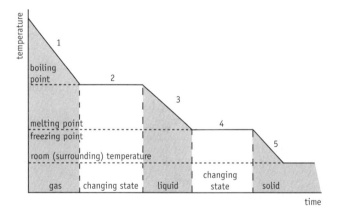

✓ Quick check 3

Quick check questions

1 Construct a table which summarizes how the kinetic energy and distance between particles change as a solid changes to a liquid and a gas
2 Account for the fact that gases diffuse.
3 Sketch out the typical cooling curve and label it fully from memory.

Structures

The **physical properties** of virtually every compound can be explained by examining its structure and bonding. Make sure you know the details of the four structures listed below. You must be able to name examples of each structure and give reasons for their particular properties.

1 Giant ionic lattices

Description: positive metal and negative non-metal ions held together in a **giant ionic lattice** by **strong electrostatic forces** of attraction.

Examples: MgO, NaCl, CsCl ▶

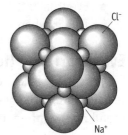

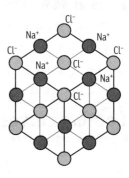

Properties	Reasons
high melting points	strong forces of attraction between ions
conduct electricity when molten or aqueous	ions are free to move and carry electrical charge
will not conduct electricity when solid	ions are held together tightly so cannot transfer energy by vibration and there are no free electrons
will not conduct heat	
dissolve in polar solvents like water	the solvent attracts ions from the lattice and surrounds them
are brittle	ion layers can become lined up with oppositely charged ions adjacent, which causes repulsion and cleavage of the crystal

2 Giant metallic structures

Description: a **lattice of positive ions** held together by a **sea of free delocalised electrons**. (This is a metallic bond.)

Examples: Al, Cu, Zn, Mg, Fe

> Remember that all that glitters isn't gold, but at least it contains free electrons...

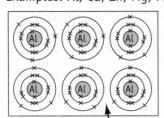
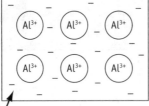
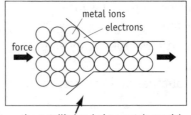

electrons in the outer shell are delocalised to form a sea of electrons
The metallic bond

the metallic bond gives metals special properties – here metal is drawn into a wire.

Properties	Reasons
high melting and boiling points	metal ions are held in position by strong bonds
good conductors of electricity	delocalised electrons are mobile, so electric charge is carried
good conductors of heat	metal ions vibrate and pass on their kinetic energy gained from heat, as do the mobile delocalised electrons
malleable and ductile	metal ions can slide past each other but still maintain the lattice structure
strong	metallic lattice is a strong design

3 Simple molecular crystals

Description: **molecules** are held together by **weak intermolecular forces** (like VDW).

Example: iodine (I_2) ▶▶

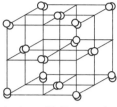
structure of iodine crystals

Properties	Reasons
low melting points	only weak intermolecular forces hold the structure together
do not conduct electricity	bonding electrons are in covalent bonds so are not free to move and carry charge
do not conduct heat	bonding electrons are in covalent bonds so are not free to move and pass on vibrations
insoluble in polar solvents like water	polar molecules will not attract the molecules unless the molecule is very polar (e.g. sugars)

4 Giant molecular crystals (macromolecular crystals)

Description: a **regular arrangement** of **atoms** joined together by **covalent bonds**.

Examples: diamond and graphite ▶▶

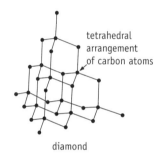
diamond — tetrahedral arrangement of carbon atoms

Properties	Reasons
high melting and boiling points	many strong covalent bonds hold the structure together
do not conduct electricity	bonding electrons are in covalent bonds so are not free to move and so conduct
do not conduct heat	bonding electrons are in covalent bonds so are not free to move and pass on vibrations
insoluble in polar solvents like water	polar molecules will not attract the atoms in the giant molecule

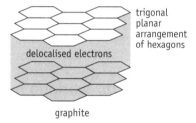
graphite — trigonal planar arrangement of hexagons, delocalised electrons

Graphite has a layered structure which allows delocalised electrons to flow between the layers and allows the layers to slide over each other.
It is therefore an electrical conductor and a good lubricant.

✓ *Quick check 1, 2*
✓ *Quick check 3–5*

Quick check questions

1. Name the four different types of structures.
2. Explain why metals can conduct electricity but covalent crystals do not.
3. If a pencil is run down a squeaking door hinge it stops squeaking, as though it had been oiled. Explain why.
4. Why do molecular crystals often have low boiling points?
5. Which of the following at stp have delocalised electrons:
   a lithium, **b** iodine, **c** oxygen, **d** carbon (diamond)?

Shapes of molecules

The shape of a simple molecule can be predicted using the **electron pair repulsion theory**. Electrons in the outer orbital of an atom exist as clouds of negative charge and repel each other. This repulsion positions them as far apart from each other as possible, around the atom. If a covalent bond forms with these electrons, the bonds will have a particular orientation and direction around the central atom and the molecule will therefore have a particular shape.

> Bonding electrons whether in a single, double or co-ordinate (dative) bond, all repel equally.

Electron pair repulsion theory

The shape of a simple covalent molecule is determined by the number of pairs of electrons around the central atom. Electrons can be...

- **bonding electrons** (shared electrons)
- **lone pairs** (non-bonding or unshared electrons)

No. of bonding pairs of e⁻	Shape	Bond angles	Arrangement of atoms	Example
2	linear	180°		$BeCl_2$
3	trigonal planar	120°		BF_3
4	tetrahedral	109.5°		CH_4
5	trigonal bipyramidal	90° and 120°		PF_5
6	octahedral	90°		SF_6

Repulsion of electron pairs

Bond pairs and lone pairs repel differently. Lone pairs are slightly closer to the nucleus than bond pairs, and therefore repel other electron pairs more readily. This affects the shape of the molecule by distorting the orbitals.

> lone pair – lone pair
> lone pair – bond pair
> bond pair – bond pair
>
> Decrease in repulsion ↓

Six rules for working out shapes of molecules

✓ *Quick check 1*

The idea is to find the number of bond pairs and lone pairs then you can work out the shape. Remember to look at the position of the nuclei around the central atom not just the total number of electron pairs, as some of them could be lone pairs.

Worked examples

	CH_4	NH_3	H_2O	ICl_4^-	NF_2^+	PCl_5
1 Group of the central atom (number of e⁻)	4	5	6	7	5	5
2 For negative ions: add an e⁻. For positive ions: subtract an e⁻	not an ion	not an ion	not an ion	add 1 as it's a -1 ion	subtract 1 as it's a +1 ion	not an ion
3 Add an e⁻ for each atom bonded to the central atom	4 + 4 = 8	5 + 3 = 8	6 + 2 = 8	8 + 4 = 12	4 + 2 = 6	5 + 5 = 10
4 Number of pairs of electrons (total from **3** ÷ 2)	8/2 = 4	8/2 = 4	8/2 = 4	12/2 = 6	6/2 = 3	10/2 = 5
5 Is this the number of bonding atoms? (If not, the extra(s) = lone pairs.)	yes (4 bond pairs 0 lone pairs)	no (3 bond pairs 1 lone pair)	no (2 bond pairs 2 lone pairs)	no (4 bond pairs 2 lone pairs)	no (2 bond pairs 1 lone pair)	yes (5 bond pairs 0 lone pairs)
6 Name the shape by looking at the arrangement of nuclei around the central atom	tetrahedral	trigonal pyramidal	bent	square planar	bent	trigonal bipyramidal
Shape	109.5°	107°	105°	90°	115°	90°, 120°

The effect of lone pairs

The first three examples in the table above have a central atom with eight electrons (that is four electron pairs). Four pairs of electrons tell us that they should have a **tetrahedral** shape like methane, but this is only true if all the electrons are bond pairs, here they are not. Nitrogen (in NH_3) has one lone pair, which distorts the tetrahedral shape and repels the N–H bond pairs pushing them closer together giving a bond angle less than 109.5°. The shape of molecules is named by looking at the nuclei around the central atom, not just the electron pairs, so the shape is trigonal pyramidal with a bond angle of 107°.

Oxygen (in H_2O) has two lone pairs, which push the two bonding pairs in the O–H bond even closer together than in ammonia, reducing the bond angle to 105°. The shape is therefore described as **bent**. The first three examples (above) are top favourites for exam questions.

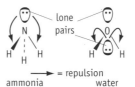

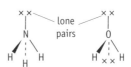

an alternative way of representing the molecules

name the shape by considering the other atoms (nuclei) around the central atom

✓ *Quick check 2, 3*

▶ The two lone pairs on oxygen repel the electrons in the O–H bonds and make the bond angle smaller than it is in ammonia

Double and triple bonds are treated like single bonds

Carbon dioxide is linear

Sulphur trioxide is trigonal planar

Sulphate ion is tetrahedral

CO_2 linear $O=C=O$

SO_3 trigonal planar

SO_4^{2-} tetrahedral

✓ *Quick check 4*

❓ Quick check questions

1. How does the repulsion between bond pairs and lone pairs differ?
2. Predict the bond angle and sketch the shape for ClF_4^-.
3. Predict the shape of Na_2BeCl_4.
4. **a** Predict the shape of BF_3 and NH_3.

   b The lone pair on the nitrogen in NH_3 can form a co-ordinate bond with the boron in BF_3. How would the bond angles in the new molecule differ from the bond angles in the original molecules?

Periodicity

The **periodic table** is the backbone of inorganic chemistry. It contains all the known elements listed in order of atomic number. Physical and chemical properties of elements change across and down the table, so we say that their properties are a **periodic function**. This just means that there's some kind of trend and the trend varies with atomic number. This regular occurring pattern is what we call **periodicity**.

The Periodic Table

Basic details of the Periodic Table

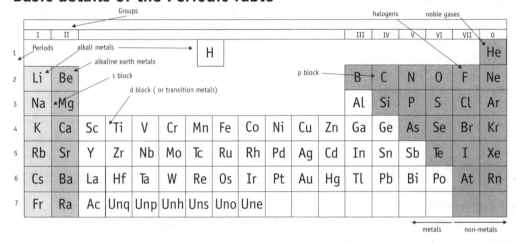

> The f block is not studied at AS or A2 level.

You need to be able to classify an element as s, p or d block and know that:
- **metals** occur to the **left** of the **zig-zag** line;
- **non-metals** occur to the **right** of the zig-zag line;
- elements **on** the zig-zag line are called **metalloids** or **semi-metals**;
- **groups** are numbered **I** to **VII** and **0**;
- **periods** are numbered **1** to **7**;
- **elements** are listed in order of **atomic number**.

✓ Quick check 1,2,3

General trends in the periodic table

	Physical	Chemical
Metals	• conductors • ductile • malleable	• reducing agents • oxides are basic in water • form cations
Non-metals	• insulators • brittle	• oxidising agents • oxides are acidic in water • form anions

Properties of elements

Atomic radius

> This is taken as half the distance between the centres of two adjacent atoms

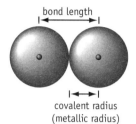
covalent radius (metallic radius)

- You may see data for covalent radius, metallic radius or van der Waals radius, depending on the bonding in question – at this level it is simplest to consider atomic radius as half the distance between centres of adjacent atoms.
- Noble gases do not bond covalently, so are usually left out when comparing atomic sizes.

Ionisation energy

> This is the amount of energy required to remove 1 mole of electrons from 1 mole of atoms in their gaseous state to form 1 mole of unipositive ions

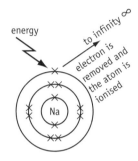

- If energy is supplied to a gaseous sample of atoms, the electrons absorb the energy and move up to higher energy levels.
- If sufficient energy is supplied the electron continues to move further and further away from the nucleus. Eventually the nucleus has no or little effect on it.
- The atom is then ionised and becomes an ion.

Electronegativity

> This is the ability of an element to attract bonding electrons in a covalent bond

- Electron clouds in a covalent bond will be distorted and move towards the more electronegative atom in the bond.
- This can cause a dipole on the molecule unless the atom is symmetrical.

✓ Quick check 2

Melting and boiling point

> The temperature at which the substance changes to a liquid or a gas

- The melting and boiling point of an element will depend on the strength of the bonds and the structure the element has.

✓ Quick check 3, 4

Quick check questions

1. Give another name for the elements in each of groups I, II, VII and 0.
2. An element X has two electrons in its outermost p sub-level.
   a describe its probable bonding and structure, b give the formula of its chloride.
3. Why is the periodic table not listed in order of mass number?
4. How is atomic radius linked to ionisation energy and electronegativity?

Module 1: Atomic structure, bonding and periodicity

Period 3 – sodium to argon

For AS level chemistry, you need to know the five trends detailed below. Make sure that you understand the reasons behind the trends and can give an explanation for patterns that you see – this is the tricky bit.

> The properties of periods are always described from left to right.

1 Atomic radius – gets smaller

From sodium to argon, electrons are being added to the third principal energy level.

- The nuclear charge is increasing from 11+ to 18+.
- The degree of shielding is the same for each successive element.
- As a result, the electrons are pulled nearer the nucleus, reducing the atomic radius.

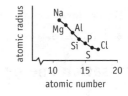

2 First ionisation energy – generally increases

The energy required to ionise the atoms generally increases across the period, because...

- the **nuclear charge** is increasing, but
- electrons are being added to the **same shell**, so the shielding is the same.

You must be able to explain the trend, especially the drop between magnesium and aluminium and between phosphorus and sulphur.

▶▶ *For more details on ionisation energy see page 11.*

3 Electronegativity – increases

This can be explained because:

- atomic radius decreases;
- nuclear charge increases.

There is therefore a greater attraction for bonded electrons.

▶▶ *For more about electronegativity see page 24.*

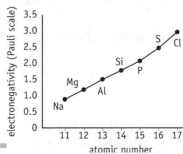

> Pure silicon is considered to be a semi-conductor, especially when it has a small amount of a group III or group V element added.

✓ *Quick check 1*

4 Electrical conductivity increases – then decreases

Conductivity increases from sodium to aluminium.

- The number of electrons in the outer shell increases.
- The number of delocalised electrons (see p. 30) per atom increases.
- There are more electrons available to 'conduct'.

Conductivity decreases from silicon to argon

- The electrons of Si, P and Cl are fixed in covalent bonds.
- Their bonding does not allow a flow of delocalised electrons.
- Particles in gases tend to be relatively very far apart, so do not conduct.

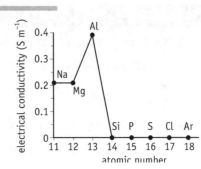

5 Melting & boiling points vary with structure and bonding

To explain this pattern you must mention

- structure – is it a giant structure or simple molecular?
- bonding – is there inter- or intramolecular bonding?

From sodium to aluminium the melting and boiling points increase.

- These are metals with giant metallic crystalline structures.
- From sodium to aluminium the number of delocalised electrons per atom increases. This increases the strength of the metallic bond which in turn increases the melting and boiling points.

Silicon has a very high melting point and boiling point.

- It is a non-metal, but it has a giant structure held together by many strong covalent bonds, similar to the tetrahedral structure of diamond.
- Each silicon atom is surrounded by four other atoms so many covalent bonds must be broken to disrupt the giant structure.

Phosphorus, sulphur and chlorine have lower melting and boiling points.

- They all have simple molecular structures held together by weak VDW forces.
- A relatively small amount of energy is enough to break these weak intermolecular surface forces.
- The surface area of the molecules decreases from phosphorus (P_4) to argon, which explains the general decrease, but sulphur forms larger rings (S_8), so will have slightly larger VDW forces, which explains the slight increase at sulphur.

Argon has the lowest melting and boiling point.

- Argon is composed of atoms which are not bonded together.
- Only weak van der Waals bonds exist between the atoms. Since the atoms are very small, these are relatively easy to break.

> It is weak intermolecular forces that are overcome when S_8, P_4 or Cl_2 turn into gases. With pure silicon, covalent bonds have to be broken, so the boiling point is higher.

✓ Quick check 2, 3

Quick check questions

1. Name two factors that affect electronegativity.
2. Explain the variation in melting and boiling points for the elements sodium to argon in terms of their structure and bonding.
3. Construct a table which illustrates at a glance how the following properties change across period 3: atomic radius, first ionisation energy, electronegativity, electrical conductivity, melting point and boiling point. *(Use an arrow pointing upwards to indicate an increase and one pointing downwards to indicate a decrease).*

Group II – alkaline earth metals (I)

You must know the physical and chemical properties of the group II elements detailed below. Like most groups in the periodic table, it's their electronic structure that determines their properties, so if you answer a question on group II, the chances are you'll need to mention electronic structure.

Physical properties of group II elements

- They are s block elements.
- They are all metals, so conduct electricity.
- Their electronic configuration ends ns^2 (where n = 1,2,3 etc.).
- They are harder than group I metals.
- Their melting points are higher than those of group I metals.
- Their metallic bonds are stronger than those of group I metals, but weaker than those of group III metals.

Group II element	Electronic configuration
Beryllium	$1s^2 2s^2$
Magnesium	$1s^2 2s^2 2p^6 3s^2$
Calcium	$1s^2 2s^2 2p^6 3s^2 3p^6 4s^2$
Barium	$[Ar]5s^2$
Strontium	$[Kr]6s^2$
Radium	$[Rn]7s^2$

Chemical properties of group II elements

- They are all reactive.
- The reactivity increases down the group.
- They produce 2+ ions.
- They have a full 2s level so are less reactive than group I metals.
- They are shiny but tarnish easily to produce a white oxide layer.

Trends you should know about

There are four specific trends that AQA say you should know about: **atomic radius, first ionisation energy, electronegativity** and **melting point**. You should also know how the metals react with **water**, and the pattern of **solubility** of the **hydroxides** and **sulphates**.

> The trends in group II are always described descending the group (going down the group).

✓ Quick check 1

1 Atomic radius – increases

- The atomic number increases, so more electrons have to be accommodated in the energy levels around the nucleus.
- Therefore more principal energy levels are needed, so the size increases.
- This increases the shielding of the outer electrons.

2 First ionisation energy – decreases

- The outer electron is in a successively higher principal energy level, which is at a greater distance from the nucleus.
- The nuclear charge increases down the group, but the inner electrons 'dilute' the effect of the positive nucleus, by shielding the outer electrons from the 'pull' of the nuclear charge.

- These two effects make it easier to remove the outer electrons going down the group.

3 Electronegativity – decreases

- The atomic radius increases going down the group so bonding pairs of electrons in a covalent bond will be further from the nucleus.
- The outer electrons are increasingly shielded from the nucleus going down the group.
- Hence the tendency for electrons to be pulled towards the nucleus is less.

> The effective nuclear charge felt by the outer electrons in all Group II elements is 2+.

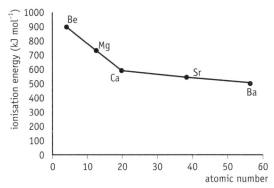

4 Melting point – decreases

- Each element has two delocalised electrons per atom but the atomic radius increases.
- This weakens the metallic bond, as it reduces the charge density on the metal ion (*just think of the electrons being 'spread thinner' over the ions, which results in a weaker metallic bond*). Hence there is a decrease in the melting point.

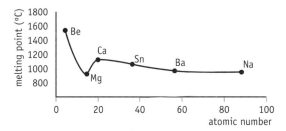

> Don't worry about magnesium. It's a bit of an oddball, as its metallic structure is different from the rest of the group.

Quick check questions

1 Name the four periodic trends which you must know about for group II. Next to each trend draw an arrow facing either up or down to show how the trend changes as the group is descended.

Group II – alkaline earth metals (II)

It's important that you know how group II elements react with water and can explain the trend observed. You must also know the trend in the solubilities of the hydroxides and sulphates (but you won't have to explain it) and be able to explain why the behaviour of beryllium is a little odd.

Group II reactions with water

You must be able to state the **products** of each reaction and explain the **trend**

- The general reactions is:

$$M(s) + 2H_2O(l) \rightarrow M(OH)_2(aq) + H_2(g)$$

But also $\quad Mg(s) + H_2O(g) \rightarrow MgO(s) + H_2(g)$

- The reaction with water becomes more vigorous as the group is descended.

Element	Reaction with water	Product
Be	Does not react with water	none
Mg	Reacts very slowly with cold water but reacts readily with steam	MgO
Ca	Reacts with cold water	$Ca(OH)_2$
Sr	Reacts with cold water	$Sr(OH)_2$
Ba	Reacts with cold water	$Ba(OH)_2$

Increasing reactivity ↓

Explanation of the trend

In the reaction, group II elements form 2+ ions – the easier it is to form the 2+ ion, the more vigorous the reaction.

- The ionisation energy of the elements decreases as the group is descended because...
  a electrons are further from the nucleus
  b the atomic radius increases
  c there is increased shielding
  ...so it takes less energy to form an ion.

- For this reason, beryllium reacts less vigorously with water than magnesium does, as it is harder for it to form the 2+ ion.

▶ This is a similar explanation to some trends in group I which you may have seen at GCSE level.

▶ Amphoteric means that the substance will react with acids and bases.

Solubility of the hydroxides – increases

- The solubility increases as the strength of the base increases.

$$M(OH)_2(s) + H_2O(l) \rightarrow M^{2+}(aq) + 2OH^-(aq)$$

- The more OH^- ions there are in the solution, the more basic the hydroxide.

Hydroxide	Solubility	Nature
$Be(OH)_2$	insoluble	amphoteric
$Mg(OH)_2$	insoluble	basic
$Ca(OH)_2$	slightly soluble	basic
$Sr(OH)_2$	soluble	basic
$Ba(OH)_2$	soluble	basic
$Ra(OH)_2$	soluble	basic

Increase in solubility of the base ↓

Solubility of the sulphates – decreases

- The larger the cation the less soluble the sulphate.

Barium sulphate – the test for the sulphate ion

Simply add a few drops of **barium chloride** (or barium nitrate) solution to the solution that may contain sulphate ions. If a white precipitate of **barium sulphate** forms – bingo...a sulphate is present.

Sulphate	Solubility
$BeSO_4$	soluble
$MgSO_4$	soluble
$CaSO_4$	slightly soluble
$SrSO_4$	insoluble
$BaSO_4$	insoluble
$RaSO_4$	insoluble

Decrease in solubility ↓

$$Ba^{2+}(aq) + SO_4^{2-}(aq) \rightarrow BaSO_4(s)$$
white ppt of barium sulphate

The solution being tested is **acidified** with dilute HCl (when using barium chloride) or dilute HNO_3 (when using barium nitrate) to prevent other precipitates forming.

✓ Quick check 1

Unusual properties of beryllium

Beryllium behaves a little differently to other members of the group. This is due to its atoms being smaller and having a higher **charge to size ratio**. All the alkaline earth metals have a 2+ charge on their ions, but with beryllium, it's spread over such a small surface area that it makes it act a bit oddly. There are three properties you must know about.

> Some books call beryllium 'atypical', which is just a different way of saying 'it's a bit unusual'.

1 $Be(OH)_2$ is amphoteric – <u>not</u> basic

Beryllium hydroxide will therefore react with both acids and bases.
- In the presence of a stronger acid it acts like a base.

$$Be(OH)_2(s) + 2HCl(aq) \rightarrow BeCl_2(aq) + 2H_2O(l)$$

- In the presence of a stronger base it acts like an acid.

$$Be(OH)_2(aq) + 2NaOH(aq) \rightarrow 2Na^+(aq) + Be(OH)_4^{2-}(aq)$$

2 $BeCl_2$ is covalent in nature – <u>not</u> ionic

- Beryllium ions are very small and have a high 2+ charge, so they have a high charge to size ratio.
- The ions are highly polarising and attract electrons on the anion.
- This means that the bonds involve 'sharing' of electrons and so are covalent in nature.
- Beryllium chloride will therefore dissolve in non-polar solvents and scarcely conducts electricity when molten.

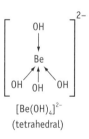
$[Be(OH)_4]^{2-}$
(tetrahedral)

3 Beryllium's maximum co-ordination number is four

- Species which have lone pairs of electrons can form co-ordinate bonds with beryllium atoms.
- The number of species surrounding beryllium is called the **co-ordination number**.
- For beryllium, this number is limited to **four** as beryllium atoms are small.
- Magnesium can form $[Mg(H_2O)_6]^{2+}$ as its atoms are larger. (The other elements in the group do not form complexes with co-ordinate bonds.)

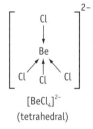
$[BeCl_4]^{2-}$
(tetrahedral)

Example
Beryllium forms co-ordinate bonds to some of the hydroxide and chloride ions to form a tetrahedral complex ion.

✓ Quick check 2, 3

Quick check questions

1. Which is more soluble: magnesium hydroxide or calcium hydroxide?
2. List three unusual properties of beryllium.
3. Lithium is found the top of Group I. Why do you think lithium iodide is soluble in propanone (an organic solvent)?

Module 1: Atomic structure, bonding and periodicity

Module 1: Exam style questions

Module 1 is all pretty mind expanding stuff, certainly not a breeze like GCSE!

To check that you follow it all, read through the module time and time again, then answer these questions. These are the kind of questions that the AQA folk will set in your module test. They'll tend to mix topics up within a module a bit more than these. They'll do this even more so on the A2 synoptic paper, which will use objective questions to test you. To answer these questions, look back to the relevant page of the module. The answers are on page 103, so you can check your understanding and mark your answers yourself. To make sure the stuff *really* goes in, try answering these questions on more than one occasion. The theory is, you should do better each time...have fun.

1 What is meant by electronic configuration? (1)

2 Why does oxygen have a first ionisation energy lower than nitrogen? (2)

3 Which one of the following would require the most energy?
   Explain your answer.

   a $Mg(g) \rightarrow Mg^+(g) + e^-$

   b $Li(g) \rightarrow Li^+(g) + e^-$

   c $Na(g) \rightarrow Na^+(g) + e^-$

   d $Al(g) \rightarrow Al^+(g) + e^-$ (2)

4 When an arrow in a box is used to represent an electron what is the significance of the arrow pointing up or down? (1)

5 Calculate the volume of hydrogen produced when 6 g of sodium hydride reacts with water. Assume the reaction occurred at stp.

   $$NaH(s) + H_2O(l) \rightarrow NaOH\ (aq) + H_2(g)$$ (4)

6 Compound X contains only boron and hydrogen. The percentage by mass of boron is 78.3%. If the largest value of m/z on a mass spectrum is 27.6, calculate the empirical formula and the molecular formula. (4)

7 Draw out the following molecules and account for their shapes.

   a H_2O and H_3O^+ (2)

   b NH_3 and NH_4^+ (2)

8 A tanker of sulphuric acid inconveniently turned over pouring 60 kg of its load into a nearby lake.

   a   Calculate the number of moles of sulphuric acid that passed into the lake. (2)

   b   It was suggested that sodium hydroxide be used to neutralise this. The equation below represents the reaction between sulphuric acid and sodium hydroxide. Balance the equation and add state symbols.

$$\text{NaOH} + \text{H}_2\text{SO}_4 \rightarrow \text{Na}_2\text{SO}_4 + \text{H}_2\text{O} \qquad (2)$$

c How many moles of sodium hydroxide would be needed to neutralise the acid? (2)

d What volume of 4 M sodium hydroxide would be needed to neutralise this many moles of acid? (2)

e Can you suggest any problems using this method of reducing the acidity? (1)

f Suggest a better method of neutralising the acid. (2)

9 A nervous student dropped a 1.5 dm³ sample of 2 M hydrochloric acid onto the laboratory floor. The student suggests using one of a number of solid substances to neutralise the acid. Calculate the mass of each substance **a → d** which would be needed to neutralise the sample exactly.

   a NaOH

   b CaO

   c Ca(OH)$_2$

   d Na$_2$CO$_3$ (12)

10 a Draw sketch graphs to represent the relationship between the following quantities for a fixed mass of an ideal gas.

x-axis	y-axis	At a constant
i volume (V)	pressure (p)	temperature (T)
ii pressure (p)	pV	temperature (T)
iii temperature (T)	pV	

(3)

b 0.50 g of a gas in a vessel of volume 1.0 dm³ exerted a pressure of 59.35 kPa at a temperature of 127°C. Use this information to calculate M_r for the gas. (4)

11 a In the production of a mass spectrum the following processes are involved:

   i vaporisation
   ii ionisation
   iii acceleration
   iv deflection
   v detection

Briefly explain what occurs in each the processes (10)

b The mass spectrum of a sample of silicon contained three peaks with mass/charge ratios and relative intensities given below:

m/z / isotope	28	29	30
Relative intensities	1	0.051	0.034

Calculate the relative atomic mass of silicon in this sample. (4)

Enthalpy change

The majority of laboratory reactions occur at atmospheric pressure, and the pressure remains constant throughout. The amount of **heat energy change** in a reaction, at constant pressure, is known as the **enthalpy change** and is given the symbol ΔH. **Enthalpy** itself cannot be measured experimentally, but it is possible to determine the enthalpy change of a reaction.

Standard enthalpy change ($\Delta H^\ominus$)

Standard enthalpy changes are those that occur under standard conditions

> Standard conditions = 100 kPa, 298 K and 1 mol dm^{-3}

- For **exothermic** reactions, $\Delta H^\ominus$ is negative as energy is lost to the system.
- For **endothermic** reactions, $\Delta H^\ominus$ is positive as energy is gained from the system.
- The **standard state** is the physical state of something under standard conditions.

There are two definitions you must know.

> A system is the environment that the reaction occurs in and remember Δ (delta) means 'change in' and the units of ΔH are kJ mol^{-1}.

1 Standard enthalpy change of formation ($\Delta H_f^\ominus$)

> The enthalpy change when 1 mole of a compound is formed from its elements in their standard states, under standard conditions of 298 K and 100 kPa.

For example:

$$C(gr) + O_2(g) \rightarrow CO_2(g) \quad \Delta H_f^\ominus = -394 \text{ kJ mol}^{-1}$$

The standard enthalpy change of formation, $\Delta H_f^\ominus$:
- for elements is zero;
- can be positive or negative;
- is illustrated by equations which must be balanced to produce 1 mole of the substance.

> When you quote a standard enthalpy change, always include an equation with state symbols. gr = graphite.

2 Standard enthalpy change of combustion ($\Delta H_c^\ominus$)

> The enthalpy change when 1 mole of a substance undergoes complete combustion in excess oxygen, under standard conditions of 298 K and 100 kPa

For example:

$$CH_4(g) + 2O_2(g) \rightarrow CO_2(g) + 2H_2O(g) \quad \Delta H_c^\ominus = -890 \text{ kJ mol}^{-1}$$

The standard enthalpy change of combustion, $\Delta H_c^\ominus$:
- is usually negative;
- is illustrated by equations which must be balanced to react 1 mole of the substance.

> Combustion means completely burned in oxygen.

Energetics

Simple calorimetry

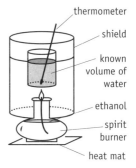

To measure the enthalpy change, the idea in calorimetry is that the energy released by burning a substance heats up some water. The temperature rise of the water is linked to the amount of energy the substance contains, but you need to know...

- the volume (and so the mass) of water;
- the mass of the substance burned.

You need to use the equation

q = heat change (kJ mol^{-1})
m = mass of water (g)
c = specific heat capacity of water (J g^{-1} K^{-1})
ΔT = temperature change (K)

$$q = mc\Delta T$$

> Specific heat capacity is the energy required to increase the temperature of 1 g of a substance by 1 K.

Bomb calorimeter – a more accurate method

Again, a known mass of substance is burned and heats a known mass of water. The difference here is that heat lost to the surroundings is massively reduced.
- The sample is electrically ignited.
- The water surrounding the calorimeter is kept at the same temperature as the water in the calorimeter by use of a heater.
- The heat capacity of the calorimeter is determined by burning a substance whose $\Delta H_c^\ominus$ is known accurately, so more accurate results may be obtained.

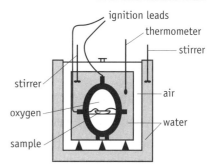

Enthalpy change for reactions in solution

In **displacement** and **neutralisation** reactions, as soon as the chemicals are mixed together any heat produced starts to be lost to the surroundings. The true temperature rise is therefore difficult to obtain. To work out the effective temperature rise, we use a graphical method, extrapolation. This means extending the range of values to follow the pattern of the graph.

> The density of water = 1 g cm^{-3}, this means that 1 cm^3 has a mass of 1 g.

✓ Quick check 3

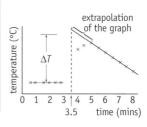

Example

For the reaction: $Zn(s) + CuSO_4(aq) \rightarrow Cu(s) + ZnSO_4(aq)$

1. Record the temperature of the CuSO$_4$ every half minute for 3 minutes.
2. At 3.5 minutes add a known mass of zinc to exactly 25.0 cm^3 of CuSO$_4$ and continue recording the temperature.
3. Extrapolate the graph back to 3.5 min to determine the temperature change. (See graph)
4. Using $q = mc\Delta T$ work out the heat change per mole of zinc.

❓ Quick check questions

1. Calculate the enthalpy of combustion of ethanol from the data in the notebook opposite.
2. Make a list of reasons why using a simple method of calorimetry does not produce data-book figures for the heat of combustion.
3. 25 cm^3 of 1 M nitric acid was placed in a polystyrene cup and neutralised with exactly 25 cm^3 of 1 M sodium hydroxide. The temperature rise for the reaction was 6.5 °C. Calculate the molar enthalpy of neutralisation for nitric acid.

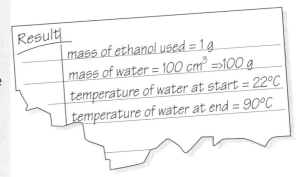

Result
mass of ethanol used = 1 g
mass of water = 100 cm^3 => 100 g
temperature of water at start = 22 °C
temperature of water at end = 90 °C

Hess's law

The **first law of thermodynamics** is all about the **conservation of energy**. It states that energy can't be created or destroyed – only changed from one form to another. Good old Hess adapted this, but said pretty much the same thing...

> The enthalpy change in a reaction depends only on the initial and final states and is independent of the route taken.

✓ Quick check 1, 2

Hess's law calculations

Consider the reaction opposite. The reactants may form the products by a **direct reaction** with enthalpy change $\Delta H_1^\ominus$. Or, they may react to make the products by an **indirect reaction**. First they form the intermediate products, with enthalpy change $\Delta H_2^\ominus$, and these in turn react to form the products of the direct reaction, with enthalpy change $\Delta H_3^\ominus$.

Hess's law states that $\Delta H_1^\ominus = \Delta H_2^\ominus + \Delta H_3^\ominus$

ΔH_1 to ΔH_3 represent the enthalpy changes for the three **reactions**.

Usually one (or more) of the enthalpy changes are difficult to determine experimentally, and therefore require a calculation to find their value.

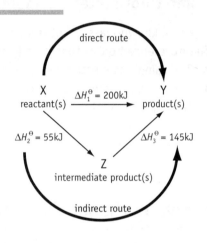

Hess's law calculations are best dealt with by drawing a **cycle** or triangle of reactions as shown opposite ▶▶

There may be more than one enthalpy value for each pathway as in the figure. Here the enthalpies are added together.

By definition $\Delta H_f^\ominus$ for elements is zero

Generally $\quad \Delta H_{(reaction)}^\ominus = \Sigma \Delta H_{f(products)}^\ominus - \Sigma \Delta H_{f(reactants)}^\ominus$

$\quad\quad\quad\quad\quad \Delta H_{(reaction)}^\ominus = \Sigma \Delta H_{c(reactants)}^\ominus - \Sigma \Delta H_{c(products)}^\ominus$

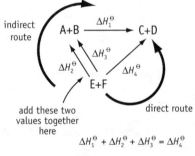

Σ = sum of

Enthalpy of formation cycles

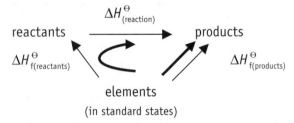

- $\Delta H_{(reaction)}^\ominus = \Sigma \Delta H_{f(products)}^\ominus - \Sigma \Delta H_{f(reactants)}^\ominus$
- You will have to rearrange the equation to find the one you need to calculate.
- The others values will probably be given in the question.

Enthalpy of combustion cycles

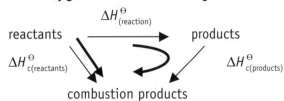

- $\Delta H_{(reaction)}^\ominus = \Sigma \Delta H_{c(reactants)}^\ominus - \Sigma \Delta H_{c(products)}^\ominus$
- You will have to rearrange the equation to find the one you need to calculate.
- The others values will probably be given in the question.

Energetics

Worked example

Find the enthalpy of formation of methane using the following data:
$\Delta H^\ominus_{c\ (carbon)} = -393.5$ kJ mol^{-1}
$\Delta H^\ominus_{c\ (hydrogen)} = -285.8$ kJ mol^{-1}
$\Delta H^\ominus_{c\ (methane)} = -890.4$ kJ mol^{-1}

Step 1 Write down the equation you are trying to find.

$$C(gr) + 2H_2(g) \longrightarrow CH_4(g)$$

> Remember:
> (g) means gas...
> (gr) means graphite.

Step 2 Assign sensible $\Delta H^\ominus$ labels to each enthalpy change (so write $\Delta H^\ominus_1$ instead of $\Delta H^\ominus_{c\ (carbon)}$ ...as it takes too long).

$\Delta H^\ominus_{f\ (methane)} = \Delta H^\ominus_1 = ?$ (unknown)
$\Delta H^\ominus_{c\ (carbon)} = \Delta H^\ominus_2 = -393.5$ kJ mol^{-1}
$\Delta H^\ominus_{c\ (hydrogen)} = \Delta H^\ominus_3 = -285.8$ kJ mol^{-1}
$\Delta H^\ominus_{c\ (methane)} = \Delta H^\ominus_4 = -890.4$ kJ mol^{-1}

Step 3 Draw a Hess's law cycle using the data given. (Put the unknown at the top.)

> There are two enthalpy values for the direct reaction ($\Delta H^\ominus_2$ and $2\Delta H^\ominus_3$), make sure you add them up.

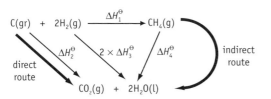

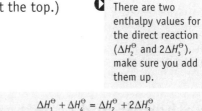

the unknown — from data — 2 × as the definition states 'per mole' and we have 2 moles of hydrogen

Follow the triangle round to find out the equation (i.e. apply Hess's law)

$$\Delta H^\ominus_1 + \Delta H^\ominus_4 = \Delta H^\ominus_2 + 2\Delta H^\ominus_3$$

Step 4 Rearrange the equation to find the unknown.
$\Delta H^\ominus_1 = \Delta H^\ominus_2 + 2\Delta H^\ominus_3 - \Delta H^\ominus_4$
$\Delta H^\ominus_1 = (-393.5) + (2 \times -285.8) - (-890.4)$
$= -74.7$ kJ mol^{-1}

> Remember that the units are kJ mol^{-1}

Step 5 Smile ☺ (optional)

✓ *Quick check 3, 4*

Quick check questions

1. State the first law of thermodynamics.
2. State Hess's law.
3. Calculate the enthalpy of reaction for the fermentation of glucose, given the following information:

   $$C_6H_{12}O_6(s) \rightarrow 2C_2H_5OH(l) + 2CO_2(g)$$

   $\Delta H^\ominus_c$(glucose) = -2820 kJ mol^{-1} and $\Delta H^\ominus_c$(ethanol) = -1367 kJ mol^{-1}
4. $\Delta H^\ominus_c$ for diamond is -397 kJ mol^{-1} and $\Delta H^\ominus_c$ for graphite is -394 kJ mol^{-1}. How much energy would be required to change 10 g of graphite into diamond?

Bond enthalpies

The **standard molar bond enthalpy** is the change when one mole of a particular bond is broken when the substance is in the gaseous phase under standard conditions. It's also known as the **standard molar enthalpy change of bond dissociation**. For example, in a diatomic molecule the electrons in the covalent bond A–B move to A and B breaking the bond, as shown below:

$$A\text{–}B(g) \rightarrow A\cdot(g) + B\cdot(g)$$

$\Delta H_{(reaction)}$ = **Bond enthalpy**

Here the covalent bond splits and each atom A and B receives one of the electrons from that formed the bond. Breaking the bond requires energy, and is represented by the equation:

$$A\text{–}B(g) \rightarrow A(g) + B(g)$$

Remember that...   ...breaking bonds requires energy   ($\Delta H^\ominus$ is positive)
                    ...making bonds releases energy   ($\Delta H^\ominus$ is negative)

▶ This process of bond breaking is known as **homolytic fission**.

Exothermic and endothermic reactions

In any reaction chemical bonds will be broken and chemical bonds will be made. Every bond will have a particular bond enthalpy.

The reaction is exothermic if...

- heat is given out to the surroundings;
- the temperature goes up;
- there is a net release of energy in the reaction;
- $\Delta H_r^\ominus$ is negative ($-\Delta H_r^\ominus$).

The reaction is endothermic if...

- heat is taken in from the surroundings;
- the temperature goes down;
- there is not a net release of energy in the reaction;
- $\Delta H_r^\ominus$ is positive ($+\Delta H_r^\ominus$).

▶ Ea = the activation energy – or energy barrier.

Worked example

Calculate the overall energy change for the following reaction

$$2H_2(g) + O_2(g) \rightarrow 2H_2O(l)$$

Bonds broken   Bonds formed

H–H   H–O–H
H–H   +   O=O   H–O–H

Step 1 Calculate the total energy needed to break the bonds (left hand side).
$2 \times$ H–H bonds = 2×436 = 872 kJ mol^{-1}

$1 \times O=O$ bonds $= 1 \times 496$ = 496 kJ mol^{-1}
Total = 1368 kJ mol^{-1}

Step 2 Calculate the total energy released making the bonds (right hand side).
$4 \times O-H$ bonds $= 4 \times 463$ = 1852 kJ mol^{-1}

Step 3 Work out the overall energy change for the reaction.

Total bonds broken − Total bonds formed
1368 − 1852 = −484 kJ mol^{-1}

> Since more energy is released forming the new bonds than is required to break the original bonds the reaction is exothermic.
>
> ✓ *Quick check 1, 2*

Mean bond enthalpy

Consider the carbon–hydrogen bonds in methane, CH_4. ▶▶

Each hydrogen is taken from a slightly different structural environment and as a result each bond enthalpy is slightly different.

- The average or mean of these would be calculated as

$$\frac{424 + 480 + 425 + 335}{4} = 416 \text{ kJ mol}^{-1}$$

Bond enthalpy	kJ mol^{-1}
$CH_4 \rightarrow CH_3 + H$	424
$CH_3 \rightarrow CH_2 + H$	480
$CH_2 \rightarrow CH + H$	425
$CH \rightarrow C + H$	335

- The mean bond enthalpy value for the C–H bond in methane is 416 kJ mol^{-1}.
- The calculation in the worked example above is only an approximation and would differ from experimental work.

Bond dissociation and mean bond enthalpy

- **Bond dissociation** refers to a particular bond in a specific compound.
- Mean bond enthalpies listed in data books are **average** values for bond enthalpies, as they consider bonds in many different compounds.

The tabulated data for a C–H bond is 413 kJ mol^{-1} which is clearly different from the average calculated for methane above.

Bond dissociation examples: O–H

$$H_2O(g) \longrightarrow H(g) + OH(g) \quad \Delta H^\ominus = +492 \text{ kJ mol}^{-1}$$

$$CH_3OH(g) \longrightarrow CH_3O(g) + H(g) \quad \Delta H^\ominus = +437 \text{ kJ mol}^{-1}$$

- Each O–H bond is in a slightly different situation regarding its nearest neighbours – this explains the difference in values.
- The mean bond enthalpy for the O–H bond (or B(O–H)) is 463 kJ mol^{-1}.

> Mean bond enthalpy is sometimes called **bond energy**.

> Remember that bond dissociation enthalpy's full and fancy title is **standard molar enthalpy change of bond dissociation**.
>
> ✓ *Quick check 3, 4*

❓ Quick check questions

1. Why do bond enthalpies have positive values?
2. Given the following values for bond energies, calculate the theoretical energy released when nitrogen reacts with hydrogen to form ammonia.

   $$N_2 + 3H_2 \rightarrow 2NH_3$$

   (N≡N = 945 kJ mol^{-1}, H–N = 391 kJ mol^{-1}, H–H = 436 kJ mol^{-1})
3. Give a definition of mean bond enthalpy.
4. Why is the C=O bond energy in CH_3CHO 736 kJ mol^{-1} but the C=O bond energy in CO_2 is 805 kJ mol^{-1}?

Rate of reaction

The **rate of a reaction** is concerned with how fast it's going. The topic of rates at AS level is called **Kinetics**, and the model you must use to explain experimental observations is called the **Collision Theory**.

> **Particle** and **species** can mean atoms, molecules or ions.

Collisions collisions collisions

Reactions occur only when an **effective collision** takes place between reacting species.

- Particles must **collide** with **sufficient energy** and in the **correct orientation**, for a reaction to occur. Many collisions occur, but few cause a reaction.
- The minimum energy with which particles need to collide to overcome the energy barrier is called the **activation energy**, E_a. E_a is different for different reactions.

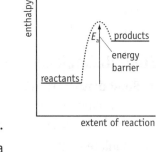

reactants must overcome the energy barrier to react

Maxwell–Boltzmann distribution curves

The distribution of **molecular energies** of a fixed mass of gas or liquid can be illustrated by the graph.▶▶

Note that if the temperature is increased from T_1 to T_2:

- The **average kinetic energy** of the molecules increases as the **temperature rises**.
- The **area** under the curve still represents the **total number of particles** present.
- The curve becomes flatter as the spread of energies increases, but the total area under the graph remains constant.

✓ *Quick check 1*

- The number of particles with very low or very high energies is small. (No molecules have zero energy and the graph meets the energy axis at energy = ∞).
- The number of particles with energy $E \geq E_a$ increases from T_1 to T_2.
- The rate of reaction increases.
- The activation energy is lowered using a catalyst because the reaction goes via a different pathway.
- More molecules have energy $E \geq E_{a(catalysed)}$.

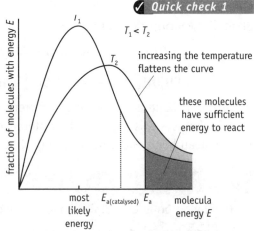

In summary: the rate of a reaction depends upon...

- the total **number of collisions**;
- the fraction of collisions with the **correct orientation**;
- the fraction of particles with **energy** $E \geq E_a$.

> A rule of thumb: the rate doubles for every 10 K rise in temperature. If a curve approaches zero at ∞, we say it **asymptotically** approaches zero.

✓ *Quick check 2*

Kinetics

To increase the rate – increase the frequency of collisions

There are five ways of increasing the frequency of collisions:

1 Increase the concentration
Low concentration
- low number of reactant particles
- collisions less likely/fewer collisions per second
- reaction less likely

High concentration
- high number of reactant particles
- collisions more likely/more collisions per second
- reaction more likely

> Light may affect some reactions, but don't worry about these kinds of reactions for the moment.

Graph: concentration of reactant [A] (mol dm^{-3}) vs time (min)
- For A + B → C
- steep curve indicates fastest part of the reaction (more A available for a collision)
- reaction slows
- when the reactant B runs out [A] remains constant

2 Increase the pressure
The effect of increasing the pressure of a gas is the same as increasing the concentration of a solution.

Low pressure
- low number of reactant particles in volume available
- collisions less likely/fewer collisions per second
- reaction less likely

High pressure
- high number of reactant particles in volume available
- collisions more likely/more collisions per second
- reaction more likely

3 Increase the surface area

Small surface area
- fewer particles available for a collision
- collisions less likely/fewer collisions per second
- reaction less likely

Large surface area
- more particles available for a collision
- collisions more likely/more collisions per second
- reaction more likely

4 Increase the temperature

Low temperature
- small amount of kinetic energy
- collisions less likely/fewer collisions per second
- reaction less likely

High temperature
- large amount of kinetic energy
- collisions more likely/more collisions per second
- reaction more likely

5 Add a suitable catalyst
A **catalyst** alters the rate of reaction, but is not used up itself.
- Addition of a catalyst reduces the activation energy ($E_{a(catalysed)} < E_{a(uncatalysed)}$).
- At the same temperature there are more particles with energy greater than or equal to $E_{a(catalysed)}$ than with energy greater than $E_{a(uncatalysed)}$.
- Some catalysts increase the rate of reaction by providing a surface for the reaction to occur on, thus increasing the number of collisions per second.

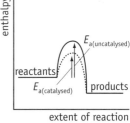

$E_{a(catalysed)} < E_{a(uncatalysed)}$

✓ *Quick check 3, 4*

Quick check questions

1. Why do collisions between particles in a gas not always cause a reaction to happen?
2. Why is the rate of reaction faster at higher temperatures?
3. List five ways of increasing the no. of collisions/sec between reacting particles.
4. What three factors determine the rate of a reaction?

Equilibria

At AS level you need to know about **reversible reactions**, which as the name suggests, don't go to completion. Instead some of the products react and change back into the reactants. It's a bit weird I know...

A + B → C + D
but then
C + D → A + B
Overall
A + B ⇌ C + D

Dynamic chemical equilibrium

When the rate of the forward reaction equals the rate of the reverse reaction, we say the reaction is in **dynamic equilibrium**.

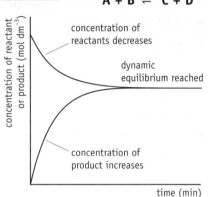

- Dynamic equilibrium is only possible in **closed systems**, where nothing is added or taken away.
- At equilibrium the forward and reverse reactions have equal **rates**.
- At equilibrium the concentrations of the reactants and products are constant.
- These reactions do not seem to be changing, e.g. there is no change in colour or change in density – we say the reaction has constant **macroscopic properties**.
- Chemical equilibria can be **homogeneous** (all the reactants are in the same **state** (phase)) or **heterogeneous** (the reactants are in different **states**).
- the **% yield** of a reaction is how much of the product is made and is defined as:

✓ *Quick check 1*

$$\% \text{ yield} = \frac{\text{Amount of product obtained}}{\text{Amount of product that could be obtained (if the reaction went to completion)}} \times 100$$

Le Chatelier's principle

The effect which conditions have on the position of equilibrium are summed up in **Le Chatelier's principle**.

> If a factor which affects the position of equilibrium is altered, the equilibrium changes and shifts in the direction which tends to reduce (and oppose) the change

A temperature change will be opposed by the equilibrium

- **increasing the temperature**: the reaction moves in the **endothermic** direction (to absorb heat).
- **decreasing the temperature**: the reaction moves in the **exothermic** direction (to release heat).

🛈 If the ΔH value is given for a reaction it is for the reaction as written from left to right.

🛈 Remember increasing the temperature will speed up the rate of reaction and so the time taken to reach equilibrium will decrease.

Equilibria

In summary

Reaction	ΔH	Temperature	[Reactants]	[Products]	Equilibrium
Exothermic	-ve	↑	↑	↓	←
Exothermic	-ve	↓	↓	↑	→
Endothermic	+ve	↑	↓	↑	→
Endothermic	+ve	↓	↑	↓	←

✓ *Quick check 2*

A concentration change will be opposed by the equilibrium

At a given temperature:

- **increasing the concentration**: the equilibrium moves to **reduce the concentration**;
- **decreasing the concentration**: the equilibrium moves to **increase the concentration**.

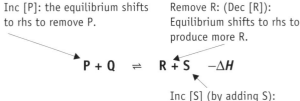

Inc [P]: the equilibrium shifts to rhs to remove P.

Remove R: (Dec [R]): Equilibrium shifts to rhs to produce more R.

$$P + Q \rightleftharpoons R + S \quad -\Delta H$$

Inc [S] (by adding S): Equilibrium shifts to lhs to reduce [S].

> [P] means concentration of P in mol dm^{-3}. This is commonly used in exams.

A pressure change will be opposed by the equilibrium

Example:

$$2V(g) + W(g) \rightleftharpoons X(g) + Y(g)$$
$$3 \text{ moles} \longrightarrow 2 \text{ moles}$$

- **increasing the total pressure**: the equilibrium shifts to the right to **reduce the pressure**;
- **decreasing the total pressure**: the equilibrium shifts to the left to **increase the pressure**.

Effect of catalysts

Catalysts have no effect on the position of equilibrium. They simply **speed up** (or some slow down) the rate at which equilibrium is achieved.

Catalysts:

- work on both the **forward** and the **reverse** reactions;
- do not alter the **composition** of the equilibrium mixture.

> The effect of pressure will depend on the reaction in question
>
> $N_2O_4(g) \rightleftharpoons 2NO_2(g)$
> 1 mole → 2 moles
> a net increase in pressure
>
> Increasing the total pressure makes the equilibrium shift to the left...but for...
>
> $H_2(g) + Cl_2(g) \rightleftharpoons 2HCl(g)$
> 2 moles → 2 moles
> no net change in pressure
>
> Increasing the total pressure of H_2 and Cl_2 will have **no effect** on the equilibrium, as there is no net increase or decrease in pressure from left to right.

Quick check questions

1. What do you understand by the term homogeneous dynamic equilibrium?
2. What effect does decreasing the temperature have on **a** the position of the equilibrium in an endothermic reaction, **b** the rate of a reaction?

Module 2: Foundation physical and inorganic chemistry

Industrial processes

There are two industrial reactions which are commonly asked about in exams. They are the **Haber process**, which makes ammonia, and the **Contact process** which makes sulphuric acid via sulphur trioxide.

The Haber process

Raw materials: Air (N_2); natural gas and water to produce H_2.

This process involves a reversible reaction which produces **ammonia** – an important raw material for the fertiliser industry.

$$N_2(g) + 3H_2(g) \rightleftharpoons 2NH_3(g) \quad \Delta H -92 \text{ kJ mol}^{-1}$$

The reaction...
- is an exothermic reaction;
- is a **homogeneous equilibrium**;
- involves a reduction in the number of moles of gas (4 moles → 2 moles), so there is a net reduction in pressure.

Optimum theoretical conditions

The optimum conditions for maximum yield are shown in the graph.
- **low temperature** – moves equilibrium to **right** increasing the yield of NH_3;
- **high pressure** – moves equilibrium to **right** increasing the yield of NH_3.

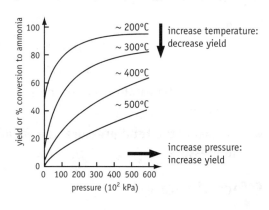

Problems
- At low temperatures the **rate of reaction is slow**, as the proportion of collisions with energy greater than or equal to the activation energy will be small.
- Industrial processes require products to be formed as quickly and as efficiently as possible, as they say **'time is money'**.
- A very high pressure would require very strong and therefore **expensive** reaction vessels – this would produce safety problems and the cost of maintaining the high pressure would be considerable.

The solution is a compromise
- The pressure used in the reaction is reduced to a more workable and less costly level.
- The temperature is raised to make the rate faster.
- A catalyst is also used to increase the rate.

Conditions:
1. temperature of 400°C
2. pressure of 200 atmospheres (20 MPa)
3. iron catalyst
4. unreacted gases are recirculated (as yield not 100%)

✓ *Quick check 1–3*

The Contact process

Raw materials: sulphur; dry air (O_2).

This is the industrial process which produces **sulphur trioxide** from sulphur dioxide. The sulphur trioxide in turn produces **sulphuric acid**. Sulphuric acid is a very important chemical as it is used in fertilisers, detergents, paints and plastics.

A summary of the reactions is given below:
Stage 1 $S(s) + O_2(g) \rightarrow SO_2(g)$
Stage 2 $2SO_2(g) + O_2(g) \rightleftharpoons 2SO_3(g)$ ΔH –197 kJ mol^{-1}
Stage 3 $SO_3(g) + H_2O(l) \rightarrow H_2SO_4(l)$

The reaction...

- is an exothermic reaction;
- is a **homogenous equilibrium**;
- involves a reduction in the moles of gas (3 moles → 2 moles), so there is a net reduction in pressure.

Optimum theoretical conditions

- **low temperature** – moves equilibrium to right increasing the yield of SO_3;
- **high pressure** – moves equilibrium to right increasing the yield of SO_3.

Problems

- The low temperature means a slow rate of reaction but a high yield – an industrial process needs as much product as possible as quickly as possible.
- The high pressure would need strong and therefore expensive reaction vessels – the cost of maintaining a high pressure would be considerable.

The solution is a compromise (again)

- The pressure used in the reaction is close to **atmospheric** as high pressure only increases the yield by about 0.5%. This saves money.
- The temperature is raised to make the rate **faster**.
- A **catalyst** is also used to increase the rate.

Conditions:

1. temperature of ~ 450°C
2. pressure of ~ 1 atmosphere (~100 kPa)
3. vanadium(V) oxide catalyst
4. gases are passed through beds of catalysts and are cooled at each stage (~97% yield).

✓ *Quick check 4*

Quick check questions

1. Why are catalysts used in industrial processes?
2. The Haber process is carried out at 20 MPa. Why is this pressure described as a compromise?
3. Why is the temperature used in the Haber process not very low?
4. What problems may arise by using the ideal reaction conditions for the Contact process?

Module 2: Foundation physical and inorganic chemistry

Redox

A redox reaction is one where **reduction** and **oxidation** occur. Reduction and oxidation can be defined in many ways.

Definitions of oxidation and reduction

Oxidation is...

- addition of oxygen
- loss of hydrogen
- loss of electrons
- an increase in oxidation state

Reduction is...

- loss of oxygen
- addition of hydrogen
- addition of electrons
- a decrease in oxidation state

Oxidation
Is
Loss (of electrons)
Reduction
Is
Gain (of electrons)

The reaction opposite can be split into **half equations** which show what is being reduced and what is being oxidised.

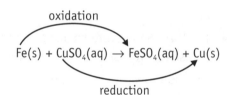
$$Fe(s) + CuSO_4(aq) \rightarrow FeSO_4(aq) + Cu(s)$$
(oxidation / reduction)

$$Fe(s) \rightarrow Fe^{2+}(aq) + 2e^-$$ Loss of electrons ∴ oxidation

$$Cu^{2+}(aq) + 2e^- \rightarrow Cu(s)$$ Gain of electrons ∴ reduction

Half equations can be combined to produce **full ionic equations**

$$Cu^{2+}(aq) + Fe(s) \rightarrow Fe^{2+}(aq) + Cu(s)$$

Oxidising and reducing agents

An **oxidising agent**...
- causes oxidation
- is an electron acceptor
- becomes reduced
- is often a non-metal

A **reducing agent**...
- causes reduction
- is an electron donor
- becomes oxidised
- is often a metal

▶ You must be able to identify which species has been reduced and which has been oxidised.

✓ *Quick check 1, 2*

Oxidation states (or oxidation numbers)

An **oxidation state**...

- describes atoms, molecules or ions;
- shows how many electrons have been used in bonding;
- can be applied to all substances, no matter what the type of bonding;
- is often given the abbreviation **Ox** (or ON).

Redox reactions

The rules for assigning oxidation states

1. For atoms of uncombined elements, the oxidation state is **zero**.
2. For compounds, the overall oxidation state is **zero**.
3. For simple ions, the oxidation state is the **charge on the ion**.
4. For complex ions – the **sum** of the oxidation states is the **charge on the ion**.
5. Covalent molecules are treated as ionic – the more **electronegative element** in the bond is assigned the more **negative oxidation** state.
6. Some elements have **variable oxidation states**, which means they have more than one oxidation state.

▶▶ *More about variable oxidation states in A2 Book*

Examples

Species	Oxidation state	Examples	Points to note
H^+ and group I ions	+1	Na^+, K^+, Ag^+	Except metal hydrides of group I, e.g. NaH Ox (Na^+) = +1 and Ox (H^-) = –1.
group II ions	+2	Mg^{2+}	
group III ions	+3	Al^{3+}	Al is always +3.
group VI ions	–2	O^{2-}, S^{2-}	Oxygen is always – 2 except in peroxides and compounds with F, e.g. H_2O_2 Ox (O) = –1.
group VII ions	–1	F^-, Cl^-, Br^-	Chlorine is –1 except in compounds with F and O where it is +ve. Fluorine is **always** –1.

Naming compounds

1. If the metal (or non-metal) has a variable oxidation state, the oxidation state is put in brackets after the metal, e.g. (Cu_2O), copper(I) oxide, (CuO) copper(II) oxide.
2. If there is no variable oxidation state don't include the number, e.g. NaI is sodium iodide not sodium(I) iodide.

Worked example

1 Calculate the oxidation number of chlorine in the ion ClO_3^-.

The Ox of chlorine in ClO_3^- is calculated by looking at the Ox of oxygen and seeing how many atoms of each element there are.

$$ClO_3^-\ (3 \times 0) + (1 \times Cl) = -1$$
$$(3 \times -2) + (1 \times Cl) = -1$$
$$-6 + Cl = -1$$
$$Cl = +5$$

> This ion is called the chlorate(V) (or *chlorate*) – "ate" indicates that oxygen is in the negative ion.

> The sum of the oxidation states = charge on the ion

Module 2: Foundation physical and inorganic chemistry

Worked example

2 Give the oxidation state of the element in bold and name the compound or element

Formula	Calculation of oxidation state	Oxidation state	Name	Comments
Na	$1 \times Na = 0$ $Na = 0$	0	sodium	For atoms in elements, the oxidation state is zero.
PCl$_3$	$(1 \times P) + (3 \times Cl) = 0$ $(1 \times P) + (3 \times -1) = 0$ $P + (-3) = 0$ $P = +3$	+3	phosphorus(III) chloride	For compounds the overall oxidation state is zero.
FeCl$_2$	$(1 \times Fe) + (2 \times Cl) = 0$ $(1 \times Fe) + (2 \times -1) = 0$ $Fe + (-2) = 0$ $Fe = +2$	+2	iron(II) chloride	Chlorine has an oxidation state of −1 so iron must be +2. Note: Iron can also be +3.
Na$_2$**Cr**$_2$O$_7$	$(2 \times Na) + (2 \times Cr) + (7 \times O) = 0$ $(2 \times +1) + (2 \times Cr) + (7 \times -2) = 0$ $+2 + (2 \times Cr) + (-14) = 0$ $2Cr + (-12) = 0$ $2\,Cr = +12$ $Cr = +6$	+6	sodium dichromate(VI)	There are two chromium atoms in the molecule. This is why the +12 is divided by 2 to give +6.
K**Mn**O$_4$	$(1 \times K) + (1 \times Mn) + (4 \times O) = 0$ $(1 \times +1) + (1 \times Mn) + (4 \times -2) = 0$ $(+1) + (1 \times Mn) + (-8) = 0$ $Mn + (-7) = 0$ $Mn = +7$	+7	potassium manganate(VII)	Manganese has an oxidation state of +7.
NO$_3^-$	$(3 \times O) + (1 \times N) = -1$ $(3 \times -2) + (1 \times N) = -1$ $-6 + N = -1$ $N = +5$	+5	nitrate(V) /nitrate	In polyatomic ions, the total oxidation state equals the charge.
NO$_2^-$	$(2 \times O) + (1 \times N) = -1$ $(2 \times -2) + (1 \times N) = -1$ $-4 + N = -1$ $N = 3$	+3	nitrate(III) /nitrite	In polyatomic ions, the total oxidation state equals the charge.
SO$_4^{2-}$	$(4 \times O) + (1 \times S) = -2$ $(4 \times -2) + (1 \times S) = -2$ $-8 + S = -2$ $S = +6$	+6	sulphate(VI) /sulphate	In polyatomic ions, the total oxidation state equals the charge.
SO$_3^{2-}$	$(3 \times O) + (1 \times S) = -2$ $(3 \times -2) + (1 \times S) = -2$ $-6 + S = -2$ $S = +4$	+4	sulphate(IV) /sulphite	In polyatomic ions, the total oxidation state equals the charge.
ClO$^-$	$(1 \times O) + (1 \times Cl) = -1$ $(1 \times -2) + (1 \times Cl) = -1$ $-2 + Cl = -1$ $Cl = +1$	+1	chlorate(I) /chlorite	Chlorine normally has an oxidation state of −1 but not with oxygen. Chlor**ate** indicates that the ion contains oxygen.
Cl$^-$	$1 \times Cl = -1$ $Cl = -1$	−1	chloride	For monoatomic ions the oxidation state equals the charge.

✓ *Quick check 3–5*

Quick check questions

1. Give the four ways of defining oxidation and reduction.
2. Explain what the difference is between an oxidising agent and a reducing agent.
3. Give the formula of **a** phosphorus(V) oxide, **b** chloric(V) acid.
4. What is the oxidation state of the element in bold: **a** **Al**Cl$_3$, **b** **H**$_2$O, **c** K**H**, **d** H**Cl**O$_2$?
5. What is the oxidation state of nitrogen in **a** NH$_4^+$, **b** NO, **c** N$_2$O, **d** N$_2$O$_4$?

Redox reactions

Redox reaction can be broken down into **half equations**. These illustrate what's happening to one element in the reaction by taking out the irrelevant bits. Half equations can be constructed using a set of rules which identify the oxidation and reduction processes. You need to be able to write half equations and be able to identify exactly the oxidation and reduction processes from specified reactants and products.

Constructing simple half equations

You will be asked to write out the half equations for given reactions. The difficult bit is ensuring that the number of electrons and the overall charge on the ions cancel out, so that the equation is balanced.

> At AS level, the chances are that you will only be asked about redox reactions involving s and p block elements. In A2 – they won't be so kind.

Worked example

Write out the half equations for the reaction of potassium iodide with chlorine water.

Step 1 Write out the equation and put oxidation states underneath each element.

$$2KI(aq) + Cl_2(aq) \rightarrow 2KCl(aq) + I_2(aq)$$
$$+1 \; -1 \quad\quad 0 \quad\quad +1 \; -1 \quad\quad 0$$

Step 2 Write out an ionic equation to show what is oxidised and reduced.

(Potassium remains unchanged so is not included.)

$$2I^-(aq) + Cl_2(aq) \rightarrow 2Cl^-(aq) + I_2(aq)$$
$$-1 \quad\quad 0 \quad\quad -1 \quad\quad 0$$

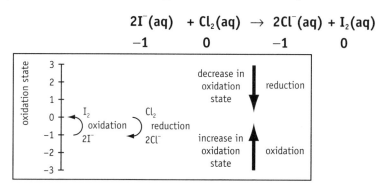

Step 3 Identify reactants and products.

$$2I^-(aq) \rightarrow I_2(aq)$$
$$Cl_2(aq) \rightarrow 2Cl^-(aq)$$

Step 4 Write out the half equations involved, by adding the correct number of electrons and identify the reaction as oxidation or reduction.

$2I^-(aq) \rightarrow I_2(aq) + 2e^-$ oxidation – (loss of electrons/increase in Ox $-1 \rightarrow 0$)

$Cl_2(aq) + 2e^- \rightarrow 2Cl^-(aq)$ reduction – (gain of electrons/decrease in Ox $0 \rightarrow -1$)

✓ *Quick check 1, 2*

Module 2: Foundation physical and inorganic chemistry

Constructing more complex half equations

The process gets just that bit more difficult with complicated ions.

Constructing complex half equations: The rules of the game...

Step 1 Write the formulae for the reactants and products and balance the atoms undergoing redox.

Step 2 Balance any oxygen atoms by adding H_2O.

Step 3 Balance any hydrogen atoms by adding H^+ ions.

Step 4 Balance the charges by adding electrons.

Step 5 Double check that the numbers of atoms are balanced.

Step 6 Double check that the numbers of charges are balanced. (The total number of electrons transferred will be equal to the total change in oxidation state.)

Step 7 Add state symbols.

Worked example

Write a half equation for the conversion of an iodate(V) ion (IO_3^-) to iodine (I_2).

Step 1 Write the formula of the reactant and product.

$$IO_3^- \rightarrow I_2$$

Step 2 Balance any oxygen atoms by adding H_2O.
There are three atoms of oxygen on the LHS so add three molecules of water to the RHS to balance the oxygen atoms.

$$IO_3^- \rightarrow I_2 + 3H_2O$$

Step 3 Balance any hydrogen atoms, by adding H^+ ions.
If we add $6H^+$ ions to the LHS these will balance.

$$IO_3^- + 6H^+ \rightarrow I_2 + 3H_2O$$

Step 4 Balance the charges by adding electrons.
We have introduced six positive charges ($6H^+$) and there is a 1− charge on the iodate(V) ion. This is now balanced by adding 5 electrons which carry a 5− charge.

$$IO_3^- + 6H^+ + 5e^- \rightarrow I_2 + 3H_2O$$

Step 5 Double check that the numbers of atoms are balanced.
There are more iodines on the RHS that the left.

$$2IO_3^- + 12H^+ + 10e^- \rightarrow I_2 + 6H_2O \text{ balanced} \quad \checkmark$$

Step 6 Double check that the numbers of charges are balanced.

The numbers of charges on each side are balanced. (+12 − 2 − 10 = 0) $\checkmark$

Double check: the iodine has been reduced (electron gain) from an oxidation state of +5 to an oxidation state of 0. This represents an overall change in oxidation states of 5, and there are two iodines to reduce. Hence ten electrons have been added.

Step 7 Add state symbols.

The half equation is: $2IO_3^-$ (aq) + $12H^+$ (aq) + $10e^- \rightarrow I_2$ (aq) + $6H_2O$ (l)

Combining half equations

Half equations for the oxidising agent and the reducing agent combine to produce an overall equation for the redox reaction.

Combining half equations: The rules of the game...

Step 1 Write one half equation showing oxidation i.e. one species losing electrons.

Step 2 Write another half equation showing reduction i.e. one species gaining electrons.

Step 3 Make sure that the number of electrons being lost and gained is the same.

Step 4 Add the two half equations together to produce the overall equation and cancel the electrons.

Step 5 Check that the final equation is balanced for atoms.

Step 6 Check that the final equation is balanced for charges on the ions.

Worked example

Write the redox equation for the oxidation of iron(II) ions by chlorine.

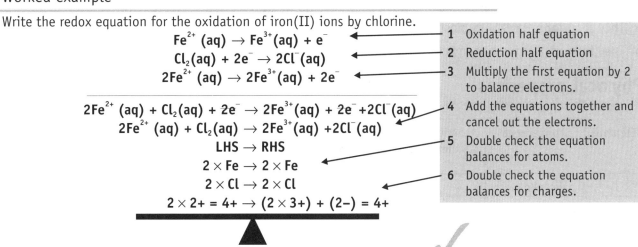

1 Oxidation half equation
$$Fe^{2+}(aq) \rightarrow Fe^{3+}(aq) + e^-$$

2 Reduction half equation
$$Cl_2(aq) + 2e^- \rightarrow 2Cl^-(aq)$$

3 Multiply the first equation by 2 to balance electrons.
$$2Fe^{2+}(aq) \rightarrow 2Fe^{3+}(aq) + 2e^-$$

4 Add the equations together and cancel out the electrons.
$$2Fe^{2+}(aq) + Cl_2(aq) + 2e^- \rightarrow 2Fe^{3+}(aq) + 2e^- + 2Cl^-(aq)$$
$$2Fe^{2+}(aq) + Cl_2(aq) \rightarrow 2Fe^{3+}(aq) + 2Cl^-(aq)$$

5 Double check the equation balances for atoms.
LHS → RHS
$2 \times Fe \rightarrow 2 \times Fe$
$2 \times Cl \rightarrow 2 \times Cl$

6 Double check the equation balances for charges.
$2 \times 2+ = 4+ \rightarrow (2 \times 3+) + (2-) = 4+$

Combined equation: $2Fe^{2+}(aq) + Cl_2(aq) \rightarrow 2Fe^{3+}(aq) + 2Cl^-(aq)$

Why combine half equations?

- They show the **real redox** processes going on.
- Electrons do not occur as separate species under normal laboratory conditions (so there is no need to show them).

✓ *Quick check 3, 4*

Quick check questions

1 Identify the oxidising and reducing species in the following reaction.
$$IO_3^-(aq) + 6H^+(aq) + 5I^-(aq) \rightarrow 3I_2(s) + 3H_2O(l)$$

2 Identify the oxidising and reducing species in the following reaction.
$$Br_2(aq) + KI(aq) \rightarrow KBr(aq) + I_2(aq)$$

3 Write a half equation for the conversion of the chlorate(V) ion to chloride.
$$ClO_3^- \rightarrow Cl^-$$

4 Combine the two half equations to give a balanced equation.
$$2S_2O_3^{2-}(aq) \rightarrow S_4O_6^{2-}(aq) + 2e^-$$
$$I_2(aq) + 2e^- \rightarrow 2I^-(aq)$$

Group VII – the halogens

Each member of group VII, the **halogens**, has an electronic configuration which ends, ns^2np^5. This gives the group its very distinctive properties and explains much of its behaviour.

General facts you should know about the halogens

- They are **non-metals** and so have general properties of non-metals.
- They are **toxic**.
- They have **coloured vapours**.
- They have molecules made up of pairs of atoms – **diatomic** molecules, e.g. F_2, Cl_2.
- They form **uninegative** ions, e.g. Cl^-, Br^-.
- They form **ionic salts** with most metals e.g. KCl, NaBr.
- They form covalent compounds with other non-metals e.g. PVC, CCl_4.

✓ *Quick check 1*

Physical properties

Element	Appearance	Atomic number	Electronic structure	Atomic radius (nm)	Electronegativity	Boiling point (°C)
F_2	pale yellow gas	9	$[He]2s^22p^5$	0.071	4.0	−188
Cl_2	pale green gas	17	$[Ne]3s^23p^5$	0.099	3.0	−35
Br_2	dark red liquid	35	$[Ar]4s^24p^5$	0.114	2.8	59
I_2	dark purple solid	53	$[Kr]5s^25p^5$	0.133	2.5	184

> Astatine, isn't included because it's radioactive and not a naturally occurring element. Fluorine is just too reactive to let you loose on it.

Atomic radius – increases

As the **atomic number increases** as the group is descended, the number of electrons per atoms increases. This means that...

- more principal energy levels are needed, as the group is descended *and*...
- as more energy levels are occupied the atom size increases.

increase in size affects many properties

Electronegativity – decreases

- Since the atoms become larger as the group is descended, the nucleus is increasingly **shielded** by the additional inner electrons.
- This increase in shielding means the effective nuclear charge is reduced, so the ability of the nucleus to attract electrons in a covalent bond is reduced.

Boiling point – increases

The stronger the intermolecular forces of attraction the higher the boiling point.

- As the molecules get bigger, the strength of the van der Waals forces between the molecules increases and so the boiling point increases.

✓ *Quick check 2–4*

Chemical properties

Halogen molecules gain electrons to form halide ions. (The molecules are reduced.)

This means that halogens act as **oxidising agents** and accept electrons.

E.g. $Br_2 + 2e^- \rightarrow 2Br^-$
Ox 0   Ox −1

The halogen's oxidising ability – decreases

- As the atomic radius increases, the effective nuclear charge felt by the peripheral electrons is reduced, due to shielding by the inner electrons. This means that the halogen atoms tend to gain electrons less readily, as the group is descended
- they therefore become **less reactive**

> Remember that an oxidising agent causes oxidation, but in doing so becomes reduced.

Displacement reactions demonstrate their oxidising power

The more reactive element displaces the less reactive element

If solutions of halogens and halide salts are mixed with small amounts of cyclohexane, the cyclohexane changes colour if a reaction occurs.

> Cyclohexane is a non-polar solvent which floats on water. Halogens look coloured when dissolved in it.

halogen \ halide solution	Cl^- (aq)	Br^- (aq)	I^- (aq)
chlorine (aq)	no change	yellow/orange solution $Cl_2(aq) + 2Br^-(aq) \rightarrow 2Cl^-(aq) + Br_2(aq)$	purple/brown solution $Cl_2(aq) + 2I^-(aq) \rightarrow 2Cl^-(aq) + I_2(s)$
bromine (aq)	no change $Br_2(aq) + 2Cl^-(aq) \rightarrow$ no reaction	no change	purple/brown solution $Br_2(aq) + 2I^-(aq) \rightarrow 2Br^-(aq) + I_2(s)$
iodine (aq)	no change $I_2(aq) + 2Cl^-(aq) \rightarrow$ no reaction	no change $I_2(aq) + 2Br^-(aq) \rightarrow$ no reaction	no change

Conclusions

- Chlorine oxidises bromide and iodide ions, so chlorine is a more powerful oxidising agent than bromine and iodine. ▶▶
- Bromine oxidises iodide ions – but not chloride ions, so it is a less powerful oxidising agent than chlorine, but a more powerful oxidising agent than iodine. ▶▶
- Iodine is not a strong enough oxidising agent to oxidise chloride or bromide ions.

The results demonstrate that chlorine is the most reactive and strongest oxidising agent of these three halogens.

$Cl_2(aq) + 2e^- \rightarrow 2Cl^-(aq)$
gain of electrons ∴ chlorine is reduced
$2Br^-(aq) \rightarrow 2e^-(aq) + Br_2(s)$
loss of electrons ∴ bromide is oxidised

$Br_2(aq) + 2e^- \rightarrow 2Br^-(aq)$
gain of electrons ∴ bromine is reduced
$2I^-(aq) \rightarrow 2e^-(aq) + I_2(s)$
loss of electrons ∴ iodide is oxidised

✓ *Quick check 5*

❓ Quick check questions

1. What is similar about the electronic structure of the halogens?
2. Account for the increase in atomic radius as group VII is descended.
3. Why is the electronegativity of chlorine less than that of fluorine?
4. Why is the boiling point of chlorine lower than that of bromine?
5. Why are iodide ions oxidised more easily than bromide ions?

Halide ions

Halide ions oxidise by losing electrons to form halogens. Halide ions are therefore good **reducing agents** as they readily become oxidised, especially the lower members of the group.

Halide ions are reducing agents

This property has the opposite trend to that shown by the halogens and is summarised below.

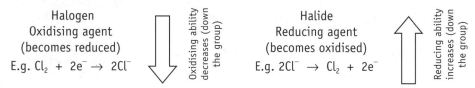

Reducing ability – increases

As the halide ions become larger in size, they become stronger **reducing agents**.
- Since the size of the halide ion increases, the outer electrons are further from the positive nuclear charge.
- The larger the ion, the easier it is for it to lose an electron.
- Iodide ions are therefore the most powerful reducing agents.
- Reducing ability therefore increases down the group.

✓ Quick check 1

Halide ions react with concentrated sulphuric acid

You need to know the different products that are formed from this reaction. They will help you understand the trend in the reducing ability of the halide ions.

▶ These reactions should be done in a fume cupboard.

Generally concentrated sulphuric acid will displace the hydrogen halide (at room temperature).

$$\text{Generally: } NaX(s) + H_2SO_4(aq) \rightarrow NaHSO_4(aq) + HX(g)$$
fumes in moist air

where X = Cl, Br or I

Sodium chloride and concentrated sulphuric acid: **$NaCl(s) + H_2SO_4(aq) \rightarrow NaHSO_4(aq) + HCl(g)$**
- Chlorine has Ox −1 in NaCl and in HCl it still has Ox −1.
- Chloride ions therefore do **not** reduce sulphuric acid.
- This is the only reaction chlorides undergo with sulphuric acid.

Sodium bromide and concentrated sulphuric acid:

Sodium bromide undergoes a redox reaction.   **$NaBr(aq) + H_2SO_4(aq) \rightarrow HBr(g) + NaHSO_4(aq)$**

Then   **$2HBr(aq) + H_2SO_4(aq) \rightarrow Br_2(g) + SO_2(g) + 2H_2O(l)$**
brown fumes colourless choking gas

- Sulphuric acid acts as an oxidising agent.
- The bromide ion acts as a reducing agent.
- Bromide ions are oxidised to bromine (Ox −1 → 0).
- Sulphur in sulphuric acid is reduced to sulphur dioxide (Ox +6 → +4).

Sodium iodide and concentrated sulphuric acid:

Iodide ion causes further reduction of the sulphur in sulphuric acid.

$NaI(s) + H_2SO_4(aq) \rightarrow HI(g) + NaHSO_4(aq)$

$2HI(aq) + H_2SO_4(aq) \rightarrow I_2(g) + SO_2(g) + 2H_2O(l)$
purple fumes

$8HI(aq) + H_2SO_4(aq) \rightarrow 4I_2(s) + H_2S(g) + 4H_2O(l)$

- Iodide ions are oxidised to iodine (Ox −1 → 0).
- Sulphur in sulphuric acid is first reduced to sulphur dioxide. Then it is further reduced to sulphur and then to the sulphide ion.

  sulphuric acid → sulphur dioxide → sulphur → hydrogen sulphide
  Ox +6 → +4 → 0 → −2
- Iodide ions are therefore stronger reducing agents than chloride and bromide ions.

✓ Quick check 2, 3

Using silver nitrate to distinguish between halide ions

If an acidified solution of silver nitrate is added to a sample of a halide salt in solution, a precipitate develops since silver halides are insoluble (except silver fluoride).

- The colour of the precipitate identifies which halide is present.
- A 'belt and braces' check to identify the halide present involves adding ammonia solution. The solubility of the precipitate confirms which halide is present.

▶ Nitric acid is used to acidify the silver nitrate. It stops other ions precipitating.

The table below shows the results you would expect.

reactant \ halide solution	F⁻(aq)	Cl⁻(aq)	Br⁻(aq)	I⁻(aq)
AgNO₃(aq)	no reaction	white precipitate	cream precipitate	yellow precipitate
Dilute NH₃(aq)	no reaction	soluble	insoluble	insoluble
Concentrated NH₃(aq)	no reaction	soluble	soluble	insoluble

Equations for the precipitation reactions

✓ Quick check 4

$AgNO_3(aq) + KCl(aq) \rightarrow AgCl(s) + KNO_3(aq)$
white ppt

$AgNO_3(aq) + KBr(aq) \rightarrow AgBr(s) + KNO_3(aq)$
cream ppt

$AgNO_3(aq) + KI(aq) \rightarrow AgI(s) + KNO_3(aq)$
yellow ppt

❓ Quick check questions

1. Briefly explain why halides are reducing agents?
2. Complete the equations:
   a $NaCl(s) + H_2SO_4(aq) \rightarrow$ b $NaBr(s) + H_2SO_4(aq) \rightarrow$
3. When sodium iodide reacts with concentrated sulphuric acid, misty fumes, a purple/black precipitate, bad eggs smell and a yellow precipitate are observed. Identify each observation.
4. How would you determine which halide ion was present in a salt?

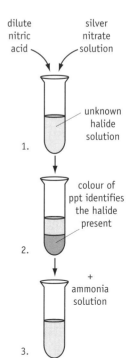

Chlorine

Chlorine is used in water treatment and the manufacture of plastics, insecticides and pesticides. This section covers some important properties of chlorine.

Chlorine and water

Chlorine and water (**chlorine water**) react to produce **chloric(I) acid**.

- Addition of alkali removes the $H^+(aq)$ from the right hand side of the above equilibrium. This makes more H_2O, which shifts the equilibrium to the right hand side and the green colour fades.
- Addition of acid adds hydrogen ions, which shifts the equilibrium to the left hand side and the green colour intensifies.
- The reaction between chlorine and water is a **redox** reaction as the chlorine has been oxidised to ClO^- and reduced to Cl^- (at the same time). This is a disproportionation reaction.

chlorine + water ⇌ hydrochloric acid + chloric(I) acid
green   colourless

$$Cl_2(g) + H_2O(l) \rightleftharpoons HCl(aq) + HClO(aq)$$
or $Cl_2(g) + H_2O(l) \rightleftharpoons 2H^+(aq) + Cl^-(aq) + ClO^-(aq)$

$$H^+(aq) + OH^-(aq) \rightarrow H_2O(l)$$

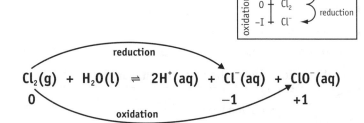

Chlorine is used in water treatment

Chloric(I) acid and sodium chlorate(I) are used as germicides to kill bacteria which live in water.

- This makes the water safe to drink, or makes the water safe to swim in.

✓ *Quick check 1, 2*

Chlorine and cold dilute sodium hydroxide

Chlorine reacts with cold dilute sodium hydroxide to produce bleach

- This reaction is a disproportionation reaction.
- Bleach is **sodium chlorate(I)** – NaClO.
- ClO^- ions act as a **bleaching agent**, which kills bacteria and oxidises dyes and stains making them colourless.

$$Cl_2(g) + 2NaOH(aq) \rightarrow NaCl(aq) + NaClO(aq) + H_2O(l)$$
or $Cl_2(g) + 2OH^-(aq) \rightarrow Cl^-(aq) + ClO^-(aq) + H_2O(l)$

> Hot NaOH(aq) produces a different set of products from cold NaOH(aq).

✓ *Quick check 3*

> The test for chlorine: if damp blue litmus is put in a gas jar of chlorine, it is turned red by the chloric(I) acid and then white, as chloric(I) acid is also a bleach.

Estimation of chlorate(I)

The amount of chlorate(I) (ClO^-) present in bleach varies, but the amount can be determined by use of sulphuric acid, potassium iodide and sodium thiosulphate.

The calculation
Worked example

10.0 cm³ of a bleach was put in a volumetric flask and made up to 250 cm³. 25.0 cm³ of this was placed in a conical flask and 1.50 g of KI was added with 10.0 cm³ of H_2SO_4. This was titrated against a standard 0.0100 M solution of sodium thiosulphate. The average of the concordant results was 24.99 cm³. Calculate the concentration of chlorine in the bleach.

Step 1 Work out the number of moles of sodium thiosulphate.

The number of moles of sodium thiosulphate used $= \dfrac{\text{concentration (moles dm}^{-3}) \times \text{volume (cm}^3)}{1000}$

$$n = cV/1000$$
$$= \dfrac{0.010 \times 24.99}{1000}$$
$$= 2.49 \times 10^{-4} \text{ mol}$$

Step 2 Work out the number of moles of iodine produced.
Using the equation:

$$I_2(aq) + 2S_2O_3^{2-}(aq) \rightarrow 2I^-(aq) + S_4O_6^{2-}(aq)$$

2 moles of $S_2O_3^{2-}$ react with 1 mole of iodine

∴ the number of moles of iodine produced $= \dfrac{2.49 \times 10^{-4}}{2} = 1.245 \times 10^{-4}$ mol

Step 3 Work out the number of moles of Cl_2.

$$Cl_2(aq) + 2I^-(aq) \rightarrow 2Cl^-(aq) + I_2(s)$$

1 mole I_2 is produced by 1 mole Cl_2

∴ 1.245×10^{-4} moles of chlorine are produced from 25.0 cm³ of the diluted bleach.

Step 4 Work out the number of moles of chlorine that would be in 250 cm³.
250 cm³ of diluted bleach has $1.245 \times 10^{-4} \times 10$ moles of chlorine as 25.0 cm³ of the 250 cm³ of diluted bleach were used.

Step 5 Work out the number of moles of chlorine in 10.0 cm³.
The number of moles of chlorine in the original 10.0 cm³ sample of bleach is also 1.245×10^{-3} moles.

Step 6 Work out the concentration in moles per dm³.
Concentration is measured in mol dm⁻³
∴ in 1000 cm³ of bleach there are $1.245 \times 10^{-3} \times 100$ moles of chlorine.
The concentration of chlorine in the bleach is <u>0.1245 mol dm⁻³</u>.

> They often make it harder to calculate the chlorine content by diluting some of the reactants.

> The calculation is summarized below

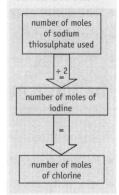

✓ *Quick check 4*

❓ Quick check questions

1. What colour is chlorine water in acid conditions?
2. What does disproportionation mean?
3. Write an equation to show the reaction of chlorine with cold sodium hydroxide solution.
4. Describe in simple terms how you would estimate the amount of chlorine in a sample of household bleach.

Reduction of metal oxides

Metal oxides are reduced to the metal using a reduction process.

The method used depends on three factors:
- the purity of metal required;
- the cost of the reducing agent used in the process;
- the energy needs of the process.

The three methods of extraction

There are three methods of extraction mentioned in the AQA specification:
- carbon reduction (e.g. iron);
- electrolysis (e.g. aluminium);
- displacement by a more reactive metal (e.g. titanium).

Iron is extracted by carbon reduction

Iron is extracted by carbon reduction in a blast furnace by a continuous process.

Raw materials: iron ore (haematite); coke; limestone; hot air.

The process:
- Coke burns in oxygen in the hot air that is blasted into the furnace.

$$C(s) + O_2(g) \rightarrow CO_2(g)$$

- Carbon dioxide then reacts with coke to produce carbon monoxide.

$$C(s) + CO_2(g) \rightarrow 2CO(g)$$

- The carbon monoxide then reduces the iron(III) oxide to iron at ~1200°C.

$$Fe_2O_3(l) + 3CO(g) \rightarrow 2Fe(l) + 3CO_2(g)$$

- In hotter parts of the furnace, carbon reduces the ore directly.

$$Fe_2O_3(l) + 3C(s) \rightarrow 2Fe(l) + 3CO(g)$$

- The molten iron flows to the bottom of the furnace and is tapped off.
- This is called 'pig iron' which is impure iron (it has ~ 4% carbon).

> Pig iron is also called cast iron. It is used to make railings, engine blocks and drain grates.

✓ *Quick check 1, 2*

Limestone removes acidic oxide impurities

- Limestone is added as it reacts with acidic oxides such as silicon dioxide.
- First the limestone decomposes to form calcium oxide at ~800°C.

$$CaCO_3(s) \rightarrow CaO(s) + CO_2(g)$$
(this is basic)

- Then calcium oxide reacts with the acidic oxides to make slag.

$$CaO(s) + SiO_2(s) \rightarrow CaSiO_3(l)$$
(slag)

- The slag floats on top of the molten iron and is tapped off.
- Slag is used in road building.

✓ *Quick check 3*

The basic oxygen converter

Pig iron contains impurities of carbon, sulphur and phosphorus, which make it brittle. The impurities are removed using the **basic oxygen converter**.

- Scrap iron is added to molten pig iron and then pure oxygen is blown through the mixture to remove the carbon and sulphur impurities. Magnesium is also added to the molten iron to help remove the sulphur.
- Lime (CaO) is added to react with any remaining solid acidic oxides, like silica and phosphorus(V) oxide forming a slag. The slag is then removed from the surface.

The carbon–iron alloys produced in a basic oxygen converter are called **steels** and typically contain between 0.04 and 1% carbon.

Limitations of carbon reduction

Carbon reduction is an effective way of obtaining metals from their ores. Carbon is cheap and readily available in the form of coke and, in theory, all metals could be produced using carbon. However, carbon reduction has a number of limitations:

- To reduce reactive metals, very high temperatures are required. Very high temperatures are uneconomic and impractical so carbon is not used with ores of reactive metals.
- It cannot reduce the ores of titanium and tungsten, as these metals form carbides.

$$\text{E.g.} \quad TiO_2(s) + 3C(s) \rightarrow TiC(s) + 2CO(g)$$

- The titanium would then have to be removed from titanium carbide, which is not an efficient way of extracting titanium.
- Sulphide ores reduced by carbon sometimes cause atmospheric pollution in the form of sulphur dioxide.
- The basic oxygen converter also produces sulphur dioxide and other acidic oxides, like phosphorus(V) oxide. These oxides can escape into the atmosphere and cause atmospheric pollution as they contribute towards acid rain.
- Carbon reduction also produces carbon monoxide and carbon dioxide which can be harmful.

▶▶ *For more on carbon monoxide and dioxide see page 82.*

✓ *Quick check 4, 5*

❓ Quick check questions

1. Name the raw materials needed to produce iron in the blast furnace.
2. Summarise the reactions that lead to the reduction of iron oxide in the blast furnace by writing balanced equations.
3. How is limestone used in the blast furnace?
4. What is the basic oxygen converter?
5. Make a list of the problems associated with using carbon to reduce metals.

Other reduction methods

The two other methods of extracting metals from their ores which you need to know about are reduction of a metal oxide by electrolysis and reduction of metal halides with a more reactive metal.

Extraction of aluminium by electrolysis

Aluminium is extracted by the electrolysis of purified aluminium oxide in a continuous process.

Raw materials: the ore (bauxite, which is mainly Al_2O_3); cryolite; lots of electricity.

The process:
- The ore is crushed and sodium hydroxide is added to remove the aluminium oxide from the ore.
- Then aluminium oxide is dissolved in molten cryolite (Na_3AlF_6) to lower its melting point by about 1000°C.
- Electricity is passed through the melt in an **electrolysis cell**.

Electrode equations
- at the cathode – aluminium forms
- at the anode – oxygen forms

$$4Al^{3+} + 12e^- \rightarrow 4Al$$
$$6O^{2-} \rightarrow 3O_2 + 12e^-$$

- the overall equation for the reaction is

$$2Al_2O_3 \rightarrow 4Al + 3O_2$$

> 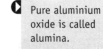 Pure aluminium oxide is called alumina.

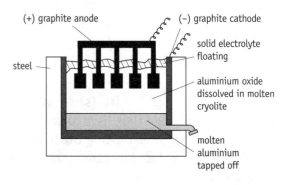

Problems with electrolysis
- The carbon anodes (at the high temperature) oxidise and need replacing.

$$2C(s) + O_2(g) \rightarrow 2CO(g)$$
and
$$C(s) + O_2(g) \rightarrow CO_2(g)$$

- Carbon monoxide and carbon dioxide gases have to be removed as carbon monoxide is poisonous and carbon dioxide is a greenhouse gas.
- Aluminium smelters require a lot of heat and electrical energy, so can only be sited near cheap electricity supplies (e.g. the HEP station near Fort William).
- Extraction of aluminium is therefore expensive.

✓ *Quick check 1*

Extraction of titanium by reduction of the metal halide

Titanium is extracted using a more reactive metal like sodium or magnesium in a batch process. The more reactive metal displaces the less reactive metal.

Titanium has a low density and a high strength and it resists corrosion – an ideal engineering material. As a result, **pure** metal is needed. Carbon reduction does not provide metal of the required purity and can render the metal quite brittle. This is of little use for most engineering purposes.

Raw materials: titanium(IV) oxide; chlorine; carbon; sodium; argon; heat.

The process:
- Ores like rutile (titanium(IV) oxide, TiO_2) are converted into titanium(IV) chloride using chlorine and carbon at 900°C.

$$TiO_2(s) + C(s) + 2Cl_2(g) \rightarrow TiCl_4(g) + CO_2(g)$$

- The titanium(IV) chloride is then reduced by sodium or magnesium in an atmosphere of argon gas at temperatures up to nearly 1000°C

$$TiCl_4(g) + 4Na(l) \rightarrow Ti(g) + 4NaCl(g)$$

- The argon prevents any air (oxygen or nitrogen) reacting with the titanium.

Problems
- Even though there is ~ 0.6% of titanium by mass in the Earth's crust the extraction process is expensive:
  1 Sodium (or magnesium) and chlorine are themselves expensive to produce.
  2 High-cost temperatures are used.
  3 An argon atmosphere has to be maintained to stop oxidation of the metal.
  4 It is a **batch process** rather than a **continuous process** – this adds to the expense of the metal as there is wasted shut down time recharging the reaction vessel.
- Chlorine is poisonous and can cause atmospheric pollution.

> Magnesium can be used to reduce the titanium(IV) chloride instead of sodium.

> Carbon reduction is definitely the cheapest of the three methods – but in the case of titanium it results in the formation of titanium carbide which contaminates the metal.

✓ *Quick check 2, 3*

Economic factors and recycling

Extraction of metals involves...
- the consumption of large amounts of energy;
- the production of a large range of gaseous by-products.

The more metal that is produced, the greater the consumption of resources and the greater the potential for pollution. One answer to this problem is the three Rs:

Recycling reduces...
- the exploitation of the finite mineral resources;
- energy demands;
- the potential for pollution (via manufacturing the metal and through waste).

Examples
- High-quality scrap metal iron is used in the manufacture of steel.
- Aluminium cans etc. are recycled and used in the electrolysis process.

> **REDUCE**
> **REUSE**
> **RECYCLE**

✓ *Quick check 4*

Quick check questions

1 Write the electrode equations for the production of aluminium and oxygen from aluminium oxide.
2 Why is an argon atmosphere used in the production of titanium?
3 Why is titanium metal expensive to buy?
4 Recycling saves 95% of the energy used to produce aluminium. Give two other benefits of recycling metals such as aluminium.

Module 2: Exam style questions

This module starts to touch on the nitty gritty chemistry which you will get to grips with if you continue with Chemistry into A2. The concepts of chemical energetics and kinetics are fundamental to many industrial reactions.

Your practical work will help you understand the concepts in this module, but you must make sure that you can link your practical work to the right bit of theory. A good way of ensuring you follow the theory is to try answering questions on it. The questions below are similar in difficulty to the ones you will find in your module tests. But watch out for questions that put *familiar chemistry* in an *unfamiliar setting*. Only practice will help you decipher the questions and get to that lovely nitty gritty chemistry. Enjoy.

1 a Define the term *standard enthalpy of formation*. (2)

   b Using the data given below, construct a Hess's law cycle to calculate the enthalpy of formation of but-1-ene (C_4H_8).

      $\Delta H_c^\ominus$ (but-1-ene) = -2716.8 kJ mol^{-1}

      $\Delta H_c^\ominus$ (hydrogen) = -285.8 kJ mol^{-1}

      $\Delta H_c^\ominus$ (carbon) = -393.5 kJ mol^{-1} (5)

2 Consider the reactions below:

    i $V(g) + W(g) \rightleftharpoons X(g) + Y(g)$ Exothermic
   ii $2V(g) + W(g) \rightleftharpoons 3X(g) + Y(g)$ Exothermic
  iii $V(g) + 2W(g) \rightleftharpoons X(g) + Y(g)$ Endothermic
  iv $V(g) + W(g) \rightleftharpoons 3X(g) + Y(g)$ Endothermic

Which reaction would increase the yield of Y if...

   a the temperature increased?

   b the pressure increased?

   c the temperature decreased and the pressure decreased?

   d the pressure decreased and the temperature increased ? (4)

3 1 g of each of the following alcohols was burned in a calorimeter. The heat evolved was used to heat 100 cm³ of water. The following temperature increases were noted.

    Methanol (CH_3OH)   52.9°C
    Ethanol (C_2H_5OH)   69.9°C
    Propan-1-ol (C_3H_7OH)   79.9°C

Calculate the enthalpy of combustion for these alcohols and predict the enthalpy of combustion for butanol, the next alcohol in the series. (7)

4 a Label the energy profile diagram opposite:▶▶

    A =
    B =
    C = (3)

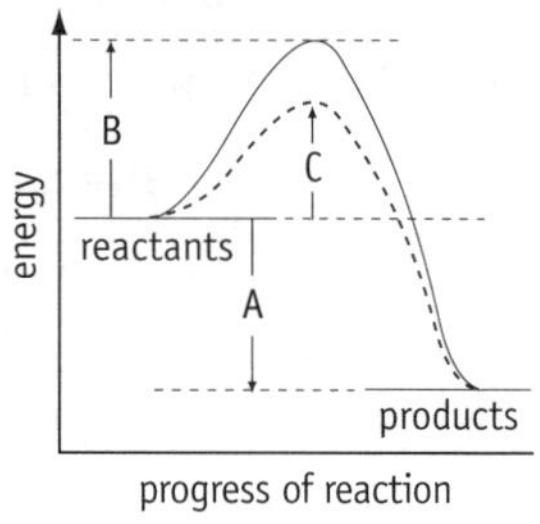

   b Give a definition of activation energy. (2)

c A small increase in temperature caused the rate of a reaction to double. Account for this observation. (2)

d Describe in terms of molecular energy how a catalyst speeds up a chemical reaction. (2)

5 a Write out the equilibrium reaction between chlorine and water. (1)

b What colour is the solution produced by this reaction? (1)

c   **i** What happens when sodium hydroxide is added to this solution? (1)
    ii Write an equation for this reaction (1)
    iii What would happen if hydrochloric acid was then added to the solution made? (1)

d What is the active ingredient of bleach? (1)

e What is the oxidation state of chlorine in this compound? (1)

6 Astatine (At) is not naturally found on Earth but is in Group VII below iodine.

a In theory, will astatine be solid, liquid or gas at room temperature and pressure? Explain your answer. (2)

b In theory hydrogen sulphide could be produced when concentrated sulphuric acid is added to potassium astatide (KAt).
  i State the role of sulphuric acid in the reaction. (1)
  ii What is the oxidation state of sulphur in sulphuric acid? (1)
  iii What is the oxidation state of sulphur in hydrogen sulphide? (1)

c Describe how you would distinguish between sodium chloride, sodium bromide and sodium iodide. (3)

7 a State three ways of extracting metals from their ores on a large scale and in each case give an equation which sums up the reaction. (6)

b In each case give two reasons why this method of extraction is used to produce large scale amounts of the metal. (6)

Naming organic compounds

There are quite literally millions of **organic** compounds, all of which contain carbon. They're found in a wide range of situations from living things to drugs, medicines and plastics. This module gives you an introduction to organic chemistry, by studying the chemistry of a number of **functional groups**.

Definitions you need to know

1 Functional group

> The group attached to the hydrocarbon chain

E.g. –OH is the **functional group** which is **substituted** for a hydrogen atom on an alkane molecule to make an alcohol. Its addition changes the behaviour of the molecule.

2 Homologous series

> A series of organic compounds with the same functional group varying only by the number of -CH$_2$ groups present

E.g. Ethanol, propanol and butanol are all in an **homologous series** called alcohols.

- Each member of the homologous series has the same **general formula**.
- Members of the group will react in the same way.
- They are often represented by R then the **functional group**, e.g. R–OH is an alcohol (R is the carbon chain).

Homologous series	Functional group
1 alkanes	–C–H
2 alkenes	>C=C<
3 haloalkanes	–X
4 alcohols	–OH
5 aldehydes	–C(=O)H
6 ketones	>C=O
7 carboxylic acids	–COOH

3 Empirical formula

> The simplest whole number ratio of the atoms present in a compound

E.g. All alkenes have the **empirical formula** CH$_2$, which means for every carbon, there are two hydrogens present – as in ethene (C$_2$H$_4$) or propene (C$_3$H$_6$).

4 Molecular formula

> The actual number of each kind of atom present in a molecule

E.g. C$_3$H$_6$ is the **molecular formula** for propene, an alkene with three carbons and six hydrogens in each molecule.

Nomenclature and isomerism

5 General formula

> Represents any member of a homologous series by giving the ratio of atoms present

E.g. Alkenes with *n* carbons in the molecule have the **general formula** – C_nH_{2n}.

✓ *Quick check 1*

6 Structural formula

> Gives information about how the carbon atoms and other groups are arranged

There are two types of structural formulae that you must know about:

a **Displayed** or **graphical formula**: shows every atom and every bond in the molecule.

b **Shorthand structural formula**: shows the same information as displayed formula but without the bonds.

E.g. Propene, C_3H_6

Displayed or graphical formula:

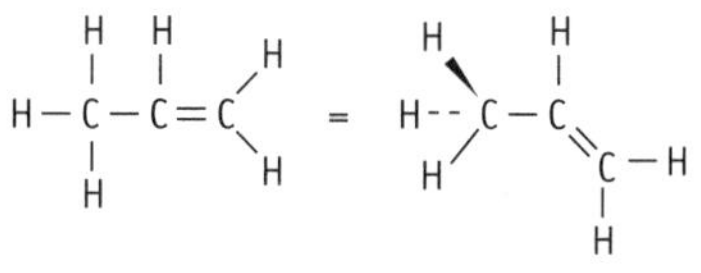

Shorthand structural formula: CH_3CHCH_2

In summary

Homologous series	General formula	Functional group	Name (suffix)	Example
1 alkanes	C_nH_{2n+2}	–C–H	-ane	C_2H_6 ethane
2 alkenes	C_nH_{2n}	C=C	-ene	C_2H_4 ethene
3 haloalkanes	$C_nH_{2n+1}X$ X is a halogen	–X	-ane (prefix: halo-)	CH_3CH_2Cl chloroethane
4 alcohols	$C_nH_{2n+2}O$	–OH	-ol (or prefix hydroxy-)	CH_3CH_2OH ethanol
5 aldehydes	$C_nH_{2n}O$	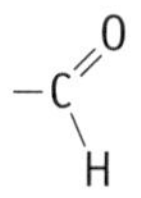 –C(=O)H	-al	CH_3CHO ethanal
6 ketones	$C_nH_{2n}O$	–C=O	-one (or prefix oxo-)	CH_3COCH_3 propanone
7 carboxylic acids	$C_nH_{2n}O_2$	–COOH	-oic acid	CH_3COOH ethanoic acid

Make sure you can recognise each of these homologous series from their functional groups.

Nomenclature – naming compounds

You must be able to recognise which family of organic chemicals a substance belongs to and apply the IUPAC rules for **systematic naming** of simple organic compounds up to six carbons long.

Names end with a suffix, names start with a prefix and infixes go within the name.

No. of carbon atoms	Prefix (or root)	
1	meth-	methyl
2	eth-	ethyl
3	prop-	propyl
4	but-	butyl
5	pent-	pentyl
6	hex-	hexyl

The general rules

1 Identify the longest parent carbon chain and assign a prefix (meth-, eth-, etc).

2 Identify the functional group(s) (e.g. –OH or –COOH) and decide on a prefix or a suffix. (*This is the group which has been substituted for a hydrogen*).

3 Number the carbons (*lowest possible*) to which the functional group(s) is attached.
(*You can number in any direction, just get the lowest possible*).

4 Name the compound by changing the name with prefixes, infixes and suffixes which describe precisely the structure of the actual compound and the position of the functional groups. (*Use commas to separate numbers and hyphens to separate numbers and letters.*)

> Anoraks see www.acdlabs.com/iupac/nomenclature for a full and exciting explanation of nomenclature.

Examples

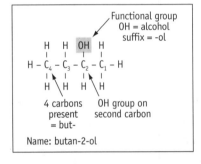

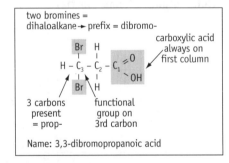

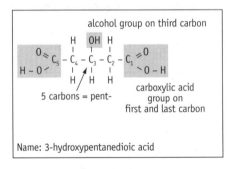

Points to note

1 Aldehydes and carboxylic acids always have the functional group on carbon 1.

2 Di or tri is added if more than one of the functional group is present.

3 If more than one functional group is present then the prefixes are in alphabetical order. When two groups would normally be suffixes, one will be the principal functional group, in the order carboxylic acid, aldehyde, ketone then alcohol as in 3-hydroxypentanedioic acid above.
There are more rules which you'll pick up in A2.

✓ *Quick check 2, 3*

Quick check questions

1 Write the general formula for alkanes, alkenes, haloalkanes, alcohols, aldehydes, ketones and carboxylic acids.

2 Name the following: **a** $CH_2CH_2CH_2CH_2CH_3$, **b** $CH_3CH(CH_3)CH_3$, **c** $CH_3CH_2CH(CH_2CH_3)CH_3$, **d** $CH_3CHBrCH_3$.

3 Give the systematic name for the anaesthetic gas fluotane, $CF_3CHClBr$.

Isomerism

Most elements form just a few compounds, but carbon forms millions. The reason is that carbon can produce many kinds of stable chains and rings. These compounds each have a particular molecular formula, but often one molecular formula can represent different arrangements of atoms in space. This gives rise to isomerism.

A general definition of isomerism is given below.

> **Isomers are compounds with the <u>same molecular</u> formulae but <u>different arrangements</u> of atoms in space**

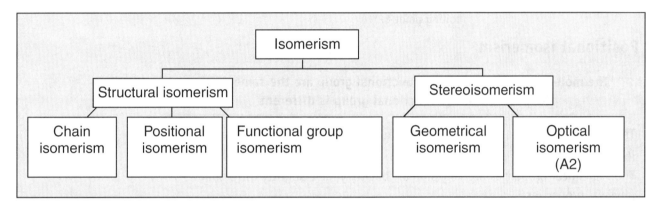

The main types of isomerism are shown in the diagram.

You need to know and understand what these terms mean, so learn the following definitions.

> Don't worry about optical isomerism until module 5 in A2 – then worry a lot.

Structural isomerism

> **The molecular formula is the same but the structural formula is different**

- The atoms present may form different bonds.
- The carbon chain may be different.
- The functional group and/or its position in the molecule may be different.

There are three types of structural isomerism you need to know about.

Module 3: Introduction to organic chemistry

1 Chain isomerism

> **The molecular formula and the functional group are the same, but the arrangement of the carbon atoms in the chain is different**

- The carbon skeleton in each isomer is different.
- Their chemical properties are very similar, but they have slightly different physical properties.

```
  H   H   H   H
  |   |   |   |
H-C - C - C - C - O - H
  |   |   |   |
  H   H   H   H
```
butan-1-ol
(boiling point −89°C)

```
  H       H
  |       |
H-C  -  C  -  C  -  O - H
  |     |     |
  H     |     H
      H-C-H
        |
        H
```
2-methylpropan-1-ol
(boiling point 82°C)

2 Positional isomerism

> **The molecular formula and the functional group are the same, but the position of the functional group is different**

- The carbon skeleton is the same, but the functional group(s) is in different positions.
- Their chemical properties are very similar, but they have slightly different physical properties.

```
  H   H   H   H   H
  |   |   |   |   |
H-C - C - C - C - C - H
  |   |   |   |   |
  Cl  Cl  H   H   H
```
1,2-dichloropentane

```
  H   H   H   H   H
  |   |   |   |   |
H-C - C - C - C - C - H
  |   |   |   |   |
  Cl  H   Cl  H   H
```
1,3-dichloropentane

- Generally the lighter an isomer is and the more branching an isomer has, the lower its boiling point.

✓ *Quick check 1, 2*

3 Functional group isomerism

> **The molecular formula is the same, but the functional group is different**

- The isomers belong to different homologous series.
- Their properties are very different.

```
  H   H   H      O
  |   |   |    //
H-C - C - C - C
  |   |   |    \
  H   H   H     H
```
butanal
(aldehyde)

```
  H   O   H   H
  |   ||  |   |
H-C - C - C - C - H
  |       |   |
  H       H   H
```
butanone
(ketone)

Carboxylic acids and esters (R-COOR') are functional group isomers as are alcohols and ethers (R-O-R'). You will learn more about these in A2.

Nomenclature and isomerism

Geometrical isomerism

> The molecular formula and the structural formula are the same, but the displayed formula is different

- The atoms form the same bonds, but they are differently arranged in space.
- It is also called **cis–trans isomerism** as *cis* and *trans* are the prefixes given to the isomers.
- '*Cis*' means 'on the same side' and '*trans*' means 'opposite' as in **trans**atlantic.

$$H-\underset{H}{\overset{H}{C}}-\overset{H}{C}=\overset{H}{C}-\underset{H}{\overset{H}{C}}-H$$

cis-but-2-ene (same side of the double bond)

trans-but-2-ene (opposite sides of the double bond)

- Geometrical isomerism is a type of **stereoisomerism**.
- Their chemical properties can be similar, but in certain situations (e.g. biological reactions) they can be very different. Physical properties like melting and boiling points are often different.

✓ *Quick check 3–5*

Bonding

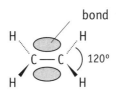

The C=C double bond will not rotate as it contains a π bond (a pi bond). This is above and below the plane of the bond, and explains why this kind of isomerism occurs in alkenes.

▶▶ *More about π bonds on page 84.*

Quick check questions

1. What is the difference between 'position isomers' and 'chain isomers'?
2. How many structural isomers does C_5H_{12} have? Name them.
3. Write out the structural formulae of the nine isomers of C_7H_{16}, and name them.
4. Explain why *cis* and *trans* isomers occur in alkenes.
5. What kind of isomers are ethanol and methoxymethane?

Module 3: Introduction to organic chemistry

Alkanes (I)

Alkanes are mainly obtained from petroleum (crude oil), which is the decayed remains of long-dead plants and animals. It is a mixture of hydrocarbons, which have different masses and surface areas and therefore have different boiling points. As a result they can be separated into fractions by **fractional distillation**.

Name	Structural formula
methane	CH_4
ethane	CH_3CH_3
propane	$CH_3CH_2CH_3$
butane	$CH_3CH_2CH_2CH_3$
pentane	$CH_3CH_2CH_2CH_2CH_3$
hexane	$CH_3CH_2CH_2CH_2CH_2CH_3$

The fractionating column

- The petroleum is heated in a furnace so it is vaporised then it passes into a **fractionating column** which is cooler at the top and hotter at the bottom.
- When the vapours reach a sufficiently cool tray, they condense and are piped off.
- The heavy, long-chain fractions, have higher boiling points, as there are more van der Waals forces acting between the molecules, so they are piped off lower down the tower.
- The lighter, shorter-chain hydrocarbons have lower boiling points, as there are less van der Waals forces acting between the molecules, so these are collected at the top of the tower.

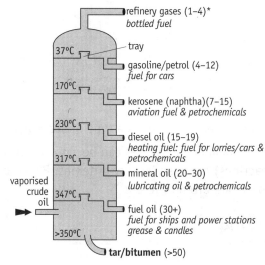

*(length of carbon chain in brackets)

✓ *Quick check 1–3*

> A fraction is a mixture of hydrocarbons with similar boiling points and similar numbers of carbons. Make sure you know the major fractions and their uses.

Cracking

- Cracking is the splitting of heavy, long-chain hydrocarbons into smaller more useful ones...like alkenes
- It involves breaking C–C and C–H bonds which requires energy.
- This is expensive, but the resulting alkenes are **high value products**.
- The alkenes can be used to make **plastics** like poly(ethene) and other chemicals.
- It meets the higher demand for the petrols and lighter fractions.

Generally

        alkanes   →   alkanes + alkenes + hydrogen
      (heavy M_r)   (light M_r) (like ethene)

There are two methods of cracking you need to know about.

> Cracking in the presence of hydrogen removes impurities of sulphur, nitrogen and some alkenes making hydrogen sulphide, ammonia and alkanes.

1 Thermal cracking – involves a free radical mechanism

- This is carried out at high temperatures of ~800°C in the absence of air.
- It is carried out at high pressures up to ~700 kPa.
- It involves homolytic fission of C–C bonds which tend to break more easily than C–H bonds (see page 48).

Bond	Energy (kJ mol^{-1})
C—C	348
C—H	416

- Radicals are produced which react with long-chain alkanes to produce more radicals in a propagation step.
- Radicals can also break C–C bonds to produce alkenes.
- Increasing the temperature tends to break the carbon–carbon bond nearer the end of the chain, which produces lower M_r alkenes.
- Thermal cracking produces a high yield of alkenes which are used as chemical feedstock.

▶▶ *See page 83 for more on free radical mechanisms.*

2 Catalytic cracking – involves a carbocation mechanism

- This is carried out at a lower temperature of ~450°C in the absence of air.
- It takes place at pressures slightly above atmospheric.
- It uses **zeolite catalysts** (a mixture of Al_2O_3 and SiO_2)
- The catalyst acts as a **Lewis acid**, which accepts an electron pair (see A2 book).
- The mechanism goes via a positive carbon – a **carbocation**.
- It produces mainly **motor fuels** and aromatic hydrocarbons which are separated by fractional distillation.

Example

$$C_8H_{18}(g) \xrightarrow{Al_2O_3/\ SiO_2} C_4H_{10}(g) + C_4H_8(g)$$

> Aromatic hydrocarbons are those which contain benzene rings (C_6H_6)

> The numbers of carbons and hydrogens on each side of the equation are equal.

✓ *Quick check 4, 5*

Complete and incomplete combustion

1. Complete combustion produces carbon dioxide and water. $CH_4(g) + 2O_2(g) \longrightarrow CO_2(g) + 2H_2O(g)$

2. Incomplete combustion produces carbon monoxide and water. $CH_4(g) + 1½O_2(g) \longrightarrow CO(g) + 2H_2O(g)$

3. Incomplete combustion produces carbon and water. $CH_4(g) + O_2(g) \longrightarrow C(s) + 2H_2O(g)$

- Carbon monoxide is a poisonous gas which bonds with the haemoglobin in red blood cells, rendering it incapable of carrying oxygen.
- To make it worse, carbon monoxide is colourless and odourless.
- Therefore badly maintained gas boilers and fires can suffocate and kill.

❓ Quick check questions

1. What is the major source of alkanes?
2. Give a use for kerosene.
3. Name a fraction obtained lower down the fractionating column than kerosene and explain why is it found lower down the column.
4. Why are thermal and catalytic cracking carried out in the absence of air?
5. What is the main reason for cracking long-chain hydrocarbons?

Alkanes (II)

Alkanes are **saturated, non-polar hydrocarbons** and unreactive. They're pretty boring really, and don't react with most reagents, like acids, bases, oxidising agents or reducing agents. You only need to know about **combustion** (see page 81), **chlorination and cracking**.

> Saturated means that they only have single bonds and non-polar means there is no major charge separation on the molecules.

Problems with complete and incomplete combustion

Combustion is an important reaction as it produces carbon dioxide and water, and releases considerable amounts of energy. Alkanes therefore make useful **fuels** and are used in the **internal combustion engine**.

The internal combustion engine produces pollutants

- The internal combustion engine is found in many motor cars.
- It produces carbon monoxide by incomplete combustion of petrol.

$$C_8H_{18} + 8½O_2 \rightarrow 8CO + 9H_2O$$

- It also produces **oxides of nitrogen** (NO_x) when the air and petrol mixture is ignited by the spark plug.

$$N_2(g) + O_2(g) \rightarrow 2NO(g)$$

- Nitrogen monoxide reacts with oxygen in the air to form nitrogen dioxide.

$$2NO(g) + O_2(g) \rightarrow 2NO_2(g)$$

- Nitrogen dioxide reacts with rain water in the atmosphere to form nitrous acid, which is further oxidised by oxygen to form nitric acid. This contributes towards acid rain.

$$4NO_2(g) + O_2(g) + 2H_2O(l) \rightarrow 4HNO_3(aq)$$

- **Unburned hydrocarbons** are also emitted.
- NO_2 reacts with oxygen or unburned hydrocarbons in sunlight to form a nasty kind of **smog**.

Catalytic converters remove pollutants

- They contain a ceramic honeycomb middle with a thin layer of platinum, palladium and rhodium.
- The metals convert the pollutant exhaust gases into harmless gases in a series of nifty reactions.

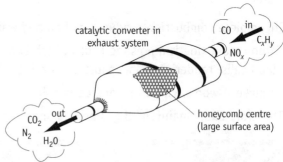

Carbon monoxide and NO_x: $2CO + 2NO \rightarrow 2CO_2 + N_2$

Unburned hydrocarbons and NO:
$C_8H_{18} + 25NO \rightarrow 8CO_2 + 12½N_2 + 9H_2O$

> Diesel engines produce solid particles of carbon, which can cause respiratory problems.

Chlorination of alkanes – free radical substitution

In simple terms, making free radicals involves breaking covalent bonds so that the bonding electrons are shared equally between the two atoms that were joined up. In AS level terms this is called **homolytic fission**. The products have unpaired electrons and go on to produce a chain reaction. The most common example AQA use is the reaction between methane and chlorine.

> Free radicals are known as intermediates, as they are formed in the reaction but are not part of the reactants or the products.
>
> ✓ *Quick check 1*

1 Initiation – UV light kick-starts the reaction

Ultra-violet light provides the energy needed to kick-start the reaction by splitting a chlorine molecule into two chlorine radicals by homolytic fission.

- The Cl–Cl bond is weaker than the C–H bonds in methane, so is broken first – this initiates the reaction.

unpaired electron of the chlorine radical

$$Cl_2 \rightarrow 2Cl\bullet$$

Bond	Bond enthalpy (kJ mol^{-1})
Cl–Cl	242
C–H (CH$_4$)	416

2 Propagation – the radicals keep it going

1. The highly reactive chlorine radicals react with the methane molecules.
2. The highly reactive methyl radicals react with the chlorine molecules.

$$CH_4 + Cl\bullet \rightarrow HCl + CH_3\bullet$$
$$CH_3\bullet + Cl_2 \rightarrow CH_3Cl + Cl\bullet$$

- The first reaction is more likely because when the H–Cl bond is formed, 431 kJ mol^{-1} of energy is released, but only 350 kJ mol^{-1} of energy is released when the C–Cl bond forms.

- Both bonds form exothermically which fuels the reaction causing a chain reaction which can be **explosive**.

- Note that a particular free radical will be used up, but others will be produced – this keeps the reaction going, until all the chlorine is used up.

> Make sure you know all the details of these three steps.

3 Termination – an end to all that radical fun

- When two radicals react they produce a more stable molecule. ▶▶
- This can lead to impurities like ethane forming.
- When all the intermediates are used up, the reaction has finished (shame ☹).

$$CH_3\bullet + Cl\bullet \rightarrow CH_3Cl$$
$$CH_3\bullet + CH_3\bullet \rightarrow CH_3CH_3$$

> Using excess methane can reduce the amount of further substitution that occurs.

Problems with this reaction

Because any or all of the hydrogen atoms on an individual methane molecule can be replaced, this reaction usually produces a mixture of all possible products. As it is a chain reaction it can also be explosive. For both these reasons it is not the best way of making chloromethane.

> ✓ *Quick check 2–4*

Quick check questions

1. Name the mechanism by which methane reacts with chlorine in UV light.
2. Name the three main steps of this mechanism.
3. Which bond breaks first to begin the reaction?
4. Why does chlorination of methane produce mixed products?

Alkenes

Alkenes are a homologous series of hydrocarbons that contain a double carbon–carbon bond (>C=C<). This makes them much more reactive than alkanes, as the double bond is 'seen' by many reactants as a high concentration of negative charge. The bond is therefore attacked by **electrophiles**, which are **electron-deficient** or **electron-seeking** species.

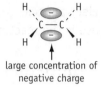

large concentration of negative charge

Facts you should know about alkenes

- They are **unsaturated** compounds as they have >C=C< double bonds
- Their general formula is C_nH_{2n}.
- They are named by looking at the alkane (parent or root) carbon chain and replacing the name ending with **-ene**.
- **Bromine water** tests for alkenes – it changes from orange to colourless.

$$CH_2=CH_2(g) \quad + \quad Br_2(aq) \quad \rightarrow \quad CH_2Br–CH_2Br$$

ethene   +   orange   →   colourless

bromine water   1,2-dibromoethane

- They undergo **electrophilic addition**.

Alkene	Formula
ethene	C_2H_4
propene	C_3H_6
butene	C_4H_8
pentene	C_5H_{10}
hexene	C_6H_{12}

> $Br_2(aq)$ can also be thought of as BrOH which will produce $CH_2(OH)CH_2Br$.

✓ *Quick check 1*

Structure and bonding

The diagram shows the structures of the first three alkenes.

- The C=C bond is a double covalent bond and contains four electrons.
- The double bond cannot rotate like the single bonds in alkanes owing to the presence of a π **bond**.
- The overlap of the single unpaired 2p electrons in each carbon forms the π bond.
- The carbon–carbon double bond and the hydrogen atoms are all in the same plane, with a bond angle of 120° (trigonal planar shape).
- This means that alkenes exhibit geometrical (*cis–trans*) isomerism, e.g. *cis*-but-2-ene and *trans*-but-2-ene.

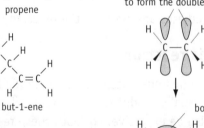

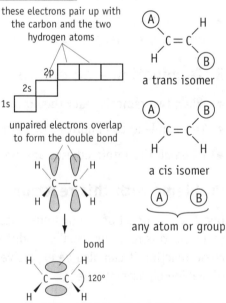

> Remember '*cis*' means 'on the same side' and '*trans*' means 'opposite' as in *trans*atlantic.

Alkenes react by electrophilic addition

Electrophiles are electron-deficient species, which seek out electron-rich species like the double bond of an alkene. This region of high concentration of electrons is approached by the electrophile and electrons are exchanged to form bonds.

Alkenes and epoxyethane

We say that the double bond is **attacked** by electrophiles to produce an **addition product**, as the electrophile becomes added to the alkene, it doesn't replace or substitute any other species.

Electrophilic addition is a two-step mechanism:

1 formation of a **carbocation**;
2 reaction with a **negative ion**.

Visualising the reaction: electrophilic addition

Step 1

- The electrophile will already be polar in nature or will have an **induced dipole** due to the concentration of negative charge in the double bond.
- The δ+ end of the electrophile begins to form a bond using electrons in the double bond. Two electrons from the double bond attack the δ+ of the electrophile. Electrons in the electrophile bond move onto the δ– atom, which breaks the bond and forms a negative ion.

electrons are repelled by the double bond and shift making the electrophile's bond polar

Step 2

- This leaves a positive charge on the carbon which is called a **carbocation**.
- The carbocation is rapidly attacked by the negative ion to complete the reaction.

Overall, the mechanism is written as:

> When X–Y split it results in an uneven share of electrons – this is known as heterolytic fission (splitting).

> Arrows show movement of a pair of electrons. They *must* go from an area of **high concentration of electrons** to an area of **low concentration**, for example from the middle of a covalent bond (or lone pair) to an electron-deficient area.

▶▶ *See pages 86 and 87 for more on electrophilic addition.*

✓ *Quick check 2–5*

❓ Quick check questions

1 List the five basic facts about alkenes that you should know.
2 Name the mechanism by which alkenes react.
3 What is an electrophile?
4 Write out a general mechanism for an electrophile attacking an alkene.
5 A known mass of two brands of margarine were separately dissolved in hexane, and bromine water was added. The first margarine took 10 drops before a brown colour was seen and the second took 15 drops. What conclusions can you draw from this?

Module 3: Introduction to organic chemistry

Reactions of alkenes (I)

The three simple alkene reactions that you always seem to see in exams are detailed below. Make sure that you know **1 the reaction conditions**, **2 the equations** and **3 the mechanism**.

Simple reactions

1 Reaction with HBr

Conditions: HBr gaseous or concentrated aqueous solution.

Equation: ethene + hydrogen bromide → bromoethane

Mechanism: ▶▶

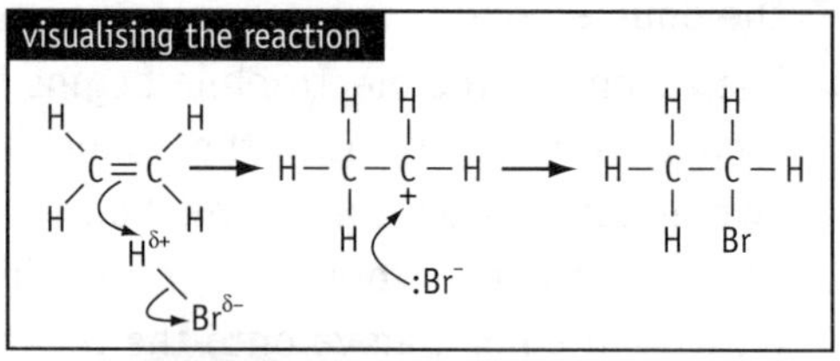

2 Reaction with Br_2

Conditions: Br_2 in water or an organic solvent.

Equation: ethene + bromine → 1,2-dibromoethane

Mechanism: ▶▶

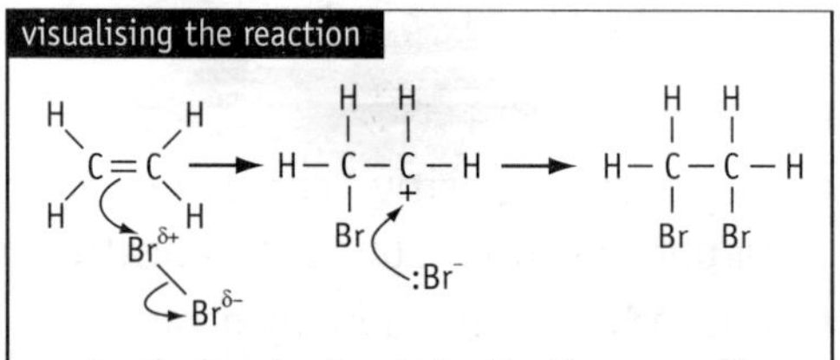

note: the bromine is polarised by the approaching electron-rich double bond

3 Reaction with H_2SO_4

Conditions: Cold concentrated sulphuric acid

Equation: ethene + sulphuric acid → ethyl hydrogensulphate

Mechanism: ▶▶

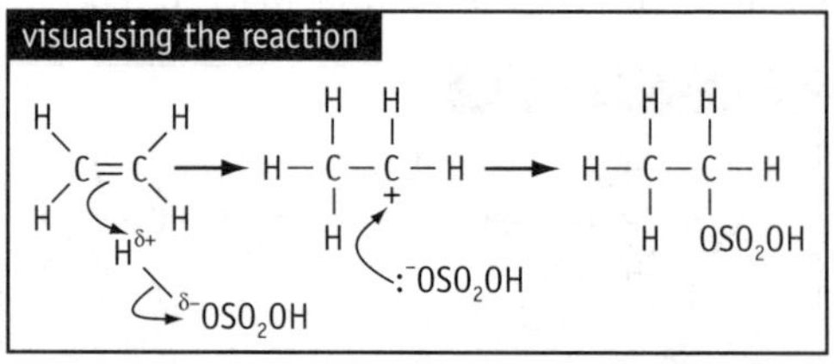

This reaction may continue...

ethyl hydrogensulphate + water → ethanol + sulphuric acid

Unsymmetrical alkenes

When propene reacts with HBr there are two possible products as shown in the diagram. ▶▶

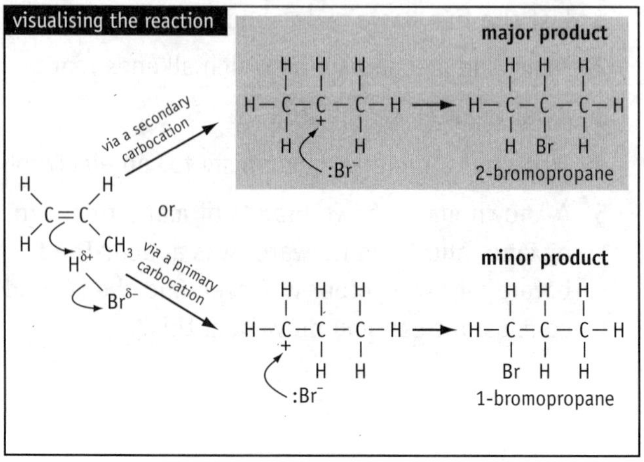

Equation for the **major product**:

$CH_3CH=CH_2$ + HBr → $CH_3CHBrCH_3$

    2-bromopropane

Equation for the **minor product**:

$CH_3CH=CH_2$ + HBr → $CH_3CH_2CH_2Br$

    1-bromopropane

Alkenes and epoxyethane

Explanation

A **secondary carbocation intermediate** (which produces the 2-bromopropane) is more stable than the primary carbocation (which produces the 1-bromopropane). This explains why there is not a 50/50 mixture of the two products.

The stability of carbocations increases **primary < secondary < tertiary**.

The reason for this is that alkyl groups (like methyl or ethyl) are **electron-releasing** relative to hydrogen, and this helps stabilise the positive charge on the carbocation.

- The more alkyl groups like -CH_3 (methyl) are attached to the carbocation intermediate, the more stable it is, as the group releases electrons onto the positive ion.
- If the ion is more stable, it will exist for a **longer time** and so will be more likely to collide with a bromide ion and react. This is why 2-bromopropane is the major product of the reaction.

one R group: $R-\overset{H}{\underset{+}{C}}-H$ primary (1°)

two R groups: $R-\overset{R'}{\underset{+}{C}}-H$ secondary (2°)

three R groups: $R-\overset{R'}{\underset{+}{C}}-R''$ tertiary (3°)

least stable ——— carbocations ——→ most stable

1° $H-\overset{CH_3 \downarrow}{\underset{+}{C}}-H$ — electrons are released by the methyl group

2° $H_3C \rightarrow \overset{CH_3 \downarrow}{\underset{+}{C}}-H$ — the more electron-releasing groups are attached to the carbocation the more stable it is

3° $H_3C \rightarrow \overset{CH \downarrow}{\underset{+}{C}} \leftarrow CH_3$

> This electron-releasing effect of the alkyl group is sometimes called the **inductive effect**.

✓ *Quick check 1*

Electrophile molecule	X(δ+)	Y(δ−)	Product with ethene	Name of product
H_2	H	H	H-C(H)(H)-C(H)(H)-H	ethane
HBr (or HCl)	H	Br (or Cl)	H-C(H)(H)-C(H)(Br)-H	bromoethane (or chloroethane)
Br_2	Br	Br	H-C(H)(Br)-C(H)(Br)-H	1,2 dibromoethane
H_2SO_4	H	HSO_4	H-C(H)(H)-C(H)(OSO_2OH)-H	ethyl hydrogensulphate

Quick check question

1 Why is a tertiary carbocation more stable than a primary carbocation?

Module 3: Introduction to organic chemistry

Reactions of alkenes (II)

Industrial reactions

1 Catalytic hydrogenation

In the presence of a catalyst, hydrogen adds across the double bond of alkenes to forms a **saturated** hydrocarbon.

$$R_2C=CR'_2 (g) + H_2(g) \xrightarrow{\text{Ni catalyst/100°C/4 atm}} R_2CH-CHR'_2$$

- The H_2 has bonded with the two carbons of the double bond.
- This reaction is used in the food industry to control the number of **unsaturated** bonds in an oil.
- It turns an oil into a fat which can be used to make margarine.
- The degree of unsaturation controls the softness.

> Some margarine adverts claim that the product is 'high in polyunsaturates' which means 'a lot of unsaturated bonds are present in the molecules'... These are better for you.

2 Hydration of alkenes to make alcohol

Ethene reacts with steam at 300°C and 7.0×10^3 kPa (70 atmospheres) with a **phosphoric acid** catalyst and produces ethanol.

$$C_2H_4 (g) + H_2O(g) \rightleftharpoons C_2H_5OH(g)$$

- This is how industrial ethanol is produced as opposed to **fermentation**.
- This is called **direct hydration** and is a continuous process (i.e. the reactants are added and the products are removed in a continuous cycle).

✓ Quick check 1

3 Polymerisation of alkenes

Alkenes have the ability to link together exothermically, usually in the presence of a catalyst, to form **addition polymers**. Ethene forms poly(ethene) (used to make plastic bowls, buckets and bin bags).

Propene forms poly(propene) (used to make climbing ropes and packaging).

$n\ C_2H_4 \longrightarrow \left(\begin{array}{cc} H & H \\ | & | \\ -C-C- \\ | & | \\ H & H \end{array} \right)_n$

many ethenes → poly(ethene)

$n\ C_3H_6 \longrightarrow \left(\begin{array}{cc} H & H \\ | & | \\ -C-C- \\ | & | \\ H & CH_3 \end{array} \right)_n$

many propenes → poly(propene)

> Always show $n \times$ the monomer going to the polymer (the repeating unit) in brackets with n after it.

✓ Quick check 2

Epoxyethane

Ethene reacts with oxygen (air) exothermically, in the presence of a silver catalyst at 180°C to make epoxyethane (pressure 1–2 MPa)

$$2CH_2=CH_2 + O_2 \xrightarrow{\text{air/Pt/180°C}} 2H_2C\overset{O}{\overset{\diagup\diagdown}{-}}CH_2$$

- Epoxyethane is an **ether** (which has a **C-O-C** bond).
- Epoxyethane is unstable, flammable and explosive, owing to the unnatural strain on the three-membered ring.

Reactions of epoxyethane

1 It reacts with water (is hydrolysed) to make ethane-1,2-diol, which is used to make antifreeze and Terylene, which is a polyester.

$$H_2C-CH_2 + H_2O \rightarrow H_2C-CH_2$$
$$\underset{O}{\diagdown \diagup} \quad\quad\quad\quad \underset{OH \;\; OH}{| \;\; |}$$

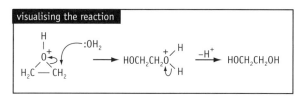

> Despite having two –OH groups ethane-1,2-diol does not get you drunk twice as fast...it is poisonous and a respiratory irritant.

> Plasticisers improve the flexibility of plastics. Surfactants are non-ionic detergents).

2 It reacts with alcohols to make alkoxyalcohols which go on to make poly(alkoxyalcohol)s

$$H_2C-CH_2 + ROH \rightarrow ROCH_2CH_2OH$$
$$\underset{O}{\diagdown \diagup}$$

$$nH_2C-CH_2 + ROH \rightarrow RO(CH_2CH_2O)_nH$$
$$\underset{O}{\diagdown \diagup}$$

Poly(alkoxyalcohol)s are used to make solvents, plasticisers and surfactants.

✓ **Quick check 3**

❓ Quick check questions

1 Propan-2-ol is made industrially from propene by hydration with steam using an acid catalyst. Write an equation for the reaction and explain why very little propan-1-ol is produced.

2 Write out an equation showing the formation of PVC poly(vinyl chloride) $-(CH_2-CHCl)_n-$.

3 Why is epoxyethane particularly reactive?

Haloalkanes

Haloalkanes (or **halogenalkanes**) have the same name as their parent alkane with a prefix for the halogen such as chloro, bromo or iodo. The halogen has simply replaced a hydrogen from the alkane structure. The presence of the electronegative halogen gives haloalkanes polar bonds. This means that they are more reactive than alkanes and are important in producing many other chemicals.

CH_3Cl

H
|
H—C—Cl
|
H

Chloromethane

Haloalkanes undergo nucleophilic substitution

- The polar nature of the bond $C^{\delta+}$–$X^{\delta-}$ results in a slightly positive carbon.
- This means that it is susceptible to attack by **nucleophiles**.
- Nucleophiles have **lone pairs** (they are electron pair donors).
- Nucleophiles attack electron-deficient carbons.
- Nucleophiles are represented by **:Nu**.

> When you say nucleophile think 'nucleus loving'... and the nucleus is positive, therefore :Nu attack δ+ carbons.

CH_3CH_2I

H H
| |
H—C—C—I
| |
H H

Iodoethane

$CH_2BrCH_2CH_2Br$

Br H Br
| | |
H—C—C—C—H
| | |
H H H

> :Nu- can be OH^-, CN^- or NH_3. Questions usually ask about primary haloalkanes (R–X).

General mechanism
1. The nucleophile attacks the $C^{\delta+}$ *(and at the same time...)*
2. the halogen leaves the molecule as a halide.

Reaction with OH⁻ (hydrolysis)

Conditions: Warm (just above room temperature) with aqueous sodium hydroxide

Equation: $CH_3CH_2Br + NaOH \rightarrow CH_3CH_2OH + NaBr$
            bromoethane   **ethanol**

Mechanism: ▶▶

visualising the reaction

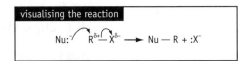

Reaction with CN⁻

Conditions: Potassium or sodium cyanide in ethanol heated under reflux.

Equation: $CH_3CH_2Br + KCN \rightarrow CH_3CH_2CN + KBr$
            bromoethane   **propanenitrile**

Mechanism: ▶▶

visualising the reaction

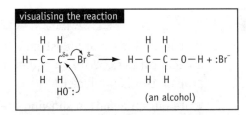
(an alcohol)

visualising the reaction

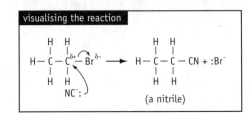
(a nitrile)

Nitriles are hydrolysed in acid to form carboxylic acids

$$CH_3CH_2CN + H_2O \xrightarrow{H^+} CH_3CH_2CONH_2$$
propanamide

$$CH_3CH_2CONH_2 + H_2O \xrightarrow{H^+} CH_3CH_2COOH + NH_3$$
propanoic acid

Haloalkanes

Reaction with NH₃

Conditions: Concentrated and excess ammonia under pressure.

Equation: $CH_3CH_2Br + 2NH_3 \rightarrow CH_3CH_2NH_2 + NH_4Br$
 bromoethane   ethylamine

Mechanism: ▶▶

Amines are important nucleophiles and can react with haloalkanes, so excess ammonia has to be added.

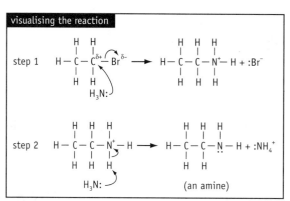

The rate of reaction increases as bond enthalpy decreases

The nucleophilic substitution mechanism involves breaking the carbon–halogen bond. The weakest carbon–halogen bond will break most readily and result in the most reactive haloalkane. Since the carbon–iodine bond is the weakest bond, iodoalkanes are more reactive than the corresponding bromo and chloroalkanes.

Mean bond enthalpy	kJ mol^{-1}
C—F	484
C—Cl	338
C—Br	276
C—I	238

Reactivity increases ↓

Elimination reactions

The OH⁻ can act as a **base** instead of a **nucleophile** which results in an **elimination reaction** producing an alkene instead of the corresponding alcohol.

Conditions: Hot sodium hydroxide in ethanol.

Equation:
$$CH_3CHBrCH_3 + NaOH \xrightarrow{ethanol} CH_3CH_2CH_2 + H_2O + NaBr$$
2-bromopropane   propene

Mechanism: ▶▶

In summary

1. The :OH⁻ attacks a hydrogen on the carbon next to the carbon–halogen bond. It forms a bond with the hydrogen atoms, using its lone pair.
2. A double bond forms and the halogen leaves as a halide ion.

Some haloalkanes produce a mixture of isomers. Thus 2-chlorobutane produces *cis*-but-2-ene, *trans*-but-2-ene and but-1-ene.

> You are unlikely to be asked about secondary haloalkanes or tertiary haloalkanes – just practise the mechanism with primary haloalkanes.

Substitution and elimination are concurrent

This means that they occur at the same time. We can increase the possibility of elimination by...

- using ethanol not water ($CH_3CH_2O^-$ is a stronger base than OH⁻);
- increasing the temperature;
- using tertiary (3°) haloalkanes (branched haloalkanes).

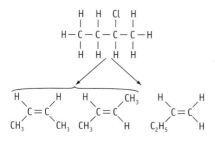

✓ Quick check 1, 2

Haloalkane	Mechanism favoured
1°	substitution
2°	substitution and elimination
3°	elimination

Quick check questions

1. Name the products when 2-bromobutane reacts with the following reagents:
   a NaOH(aq), **b** KCN, **c** NH₃, **d** NaOH$_{(ethanol)}$.
2. Write out the mechanism for elimination.

Module 3: Introduction to organic chemistry

Alcohols

Alcohols are a homologous series which contain the **–OH** functional group. They are named from the parent alkane, where one hydrogen has been replaced by the OH group.

propan-2-ol

Facts you should know about alcohols

- Their general formula is $C_nH_{2n+1}OH$.
- Ethanol is produced on an industrial scale by the **hydration** of ethene with steam and by **fermentation** of sugars.
- Alcohols are **soluble** in water because they can hydrogen bond with the water.
- Alcohol molecules are polar which leads to **hydrogen bonding**.
- Alcohols **burn** cleanly in air to give CO_2 and H_2O and so make good fuels.
- They are classified as **primary, secondary** and **tertiary** alcohols.

Name	Structural formula
methanol	CH_3OH
ethanol	CH_3CH_2OH
propan-1-ol	$CH_3CH_2CH_2OH$
propan-2-ol	$CH_3CH(OH)CH_3$

butan-1-ol	butan-2-ol	2-methyl-propan-2-ol
primary	secondary	tertiary
one R group	two R groups	three R groups
R — OH	R — C(OH)(H) — R'	R — C(R')(OH) — R''

Production of ethanol by fermentation and hydration of ethene

Ethanol is produced by two methods:

1 Fermentation: $C_6H_{12}O_6 \rightarrow 2C_2H_5OH + 2CO_2$

Conditions: 37°C, yeast, batched over several days

2 Hydration of alkenes: $C_2H_4(g) + H_2O(g) \rightleftharpoons C_2H_5OH(g)$

Conditions: 300°C and 7.0×10^3 kPa (70 atm), phosphoric acid catalyst, continuous process

Each process of making ethanol has its advantages and disadvantages.

✓ *Quick check 1*

Pros (advantages)
- Hydration is fast, produces pure ethanol and is continuous (which is cheaper).
- Fermentation uses sugars which are a renewable resource.

Cons (disadvantages)
- Hydration uses ethene which is derived from oil which is finite.
- Fermentation is slow, produces impure ethanol and is a batch process (which is expensive).

Primary alcohols are oxidised to aldehydes and carboxylic acids

1 The oxidising agent is acidified **potassium dichromate(VI)** which is represented by the symbol **[O]**.

Conditions: Warm acidified potassium dichromate (VI), aldehyde distils off.

Equation: CH_3CH_2OH + [O] → CH_3CHO + H_2O
ethanol   [oxidising agent]   ethanal   water

- The aldehyde is separated by distillation.

2 The aldehyde can be oxidised further to a carboxylic acid.

Conditions: Excess oxidising agent, acidified potassium dichromate(VI).

Equation: $CH_3CHO + [O] \rightarrow CH_3COOH + H_2O$

- The potassium dichromate(VI) changes from orange to green.
- The carboxylic acid is eventually **distilled** off after **refluxing**.

Secondary alcohols are oxidised to ketones – but not to carboxylic acids

Conditions: Warm acidified potassium dichromate(VI), ketone distils off.

Equation:

$$\begin{array}{c}H\ OH\ H\\|\ \ |\ \ |\\H-C-C-C-H\\|\ \ |\ \ |\\H\ \ H\ \ H\end{array} + [O] \longrightarrow \begin{array}{c}H\ O\ H\\|\ \ \|\ \ |\\H-C-C-C-H\\|\ \ \ \ \ |\\H\ \ \ \ \ H\end{array} + H_2O$$

propan-2-ol   [oxidising agent]   propanone   water

- The ketone is not further oxidised under normal conditions, as this would mean breaking a carbon–carbon bond.
- This fact can be used to distinguish between aldehydes and ketones.

Distinguishing between aldehydes and ketones

Aldehydes and ketones can be distinguished using Tollens reagent or Fehling's solution – two mild oxidising agents. A positive test indicates the presence of an aldehyde.

Tollens reagent with aldehydes:

    colourless   →   *gently warm*   →   silver mirror
$AgNO_3$ (in ammonia solution) →   Ag (metal)

Fehling's solution with aldehydes:

    blue   →   *gently warm*   →   brick red
(Cu^{2+} ions in alkali) →   (copper(I) oxide, Cu_2O)

- Ketones are resistant to oxidation so show a negative result with both Tollens reagent and Fehling's solution.

Reduction of aldehydes and ketones using $NaBH_4$

Aldehydes and ketones are reduced to alcohols by sodium tetrahydridoborate(III). Conditions: $NaBH_4$ in methanol (which is given the symbol [H] for simplicity).

Equations: $RCHO + 2[H] \rightarrow RCH_2OH$
primary alcohol

$RCOR' + 2[H] \rightarrow RCH(OH)R'$
secondary alcohol

> Catalytic hydrogenation using nickel or platinum will completely saturate all double bonds in aldehydes and ketones to produce alcohols.

Elimination

When alcohols are heated with concentrated sulphuric acid, water is **eliminated** to produce alkenes in a **dehydration** reaction.

Conditions: 180°C, conc. H_2SO_4 or H_3PO_4.

The 3 step mechanism:

1 **Protonation** – Concentrated sulphuric acid protonates the hydroxyl group.
2 **Loss of OH_2** – The leaving group, OH_2, is lost.
3 **Loss of hydrogen (proton)** – Hydrogen is then lost from the carbocation to form the alkene.

> Sulphuric acid is a catalyst in this reaction.

Example

✓ *Quick check 3, 4*

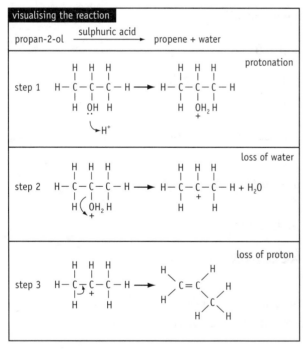

Quick check questions

1 Name two ways of producing ethanol industrially. Write equations for each of the reactions.
2 Describe how you would distinguish between an aldehyde and a ketone.
3 Describe the main steps of the elimination reaction of an alcohol.
4 Write an equation for the dehydration of cyclohexanol via elimination.

Module 3: Exam style questions

Well, are you fluent in the language of chemistry?
The problem with studying organic chemistry is that you are overwhelmed by a tidal wave of new facts, concepts and a ridiculous amount of new language.

The hardest but most important thing is absorbing all this new language. But this is the key to understanding all the facts and concepts and the way you'll show the examiner that you know your stuff. If AQA ask you *'Explain the difference between an electrophile and a nucleophile'* and you don't know what the terms *electrophile* or *nucleophile* mean – there's no way you are going to be able to answer the question – believe me. If you study French, you'll have to know the French language and it's the same with Chemistry...you must be able to speak the *language of Chemistry*.

Anyway you've probably had enough preaching...so answer these questions by looking back through the module. The answers are on page 104. Mark them to check your understanding. You'll really have to answer these on more than one occasion – to ensure the stuff really goes in. Enjoy...

1 Explain what is meant by the following terms:

   a mechanism

   b electrophile

   c nucleophile

   d free radical

   e electrophilic addition

   f nucleophilic substitution

   g elimination (7)

2 Complete the table

Functional group	Name of reaction mechanism commonly used
alkanes	
alkenes	
haloalkanes	
alcohols	

(4)

3 Methane reacts with chlorine in the presence of sunlight to produce a mixture of organic compounds.

   a Name the mechanism used in this reaction. (1)

Module 3: Introduction to organic chemistry

    b Why is sunlight needed? (1)

    c Write an overall equation to show the production of chloromethane. (1)

    d One product was found to have a relative molecular mass of 119.5. Identify this product and write an equation to show how it may have formed. (2)

    e What will be produced if two methyl radicals react? (1)

    f Why is this reaction sometimes described as a chain reaction? (1)

4 a Pentan-3-ol is oxidised by acidified potassium dichromate(VI). Using [O] to represent the oxidising agent, write an equation for this reaction and show the structure of the product. (2)

    b Pentan-3-ol reacts with concentrated sulphuric acid.

        i Name this type of reaction. (1)

        ii Give the reaction conditions. (1)

        iii Outline the mechanism. (3)

        iv Name the organic product. (1)

    c An isomer of pentanol is $(CH_3)_2C(OH)CH_2CH_3$. Describe what you would see if acidified potassium dichromate(VI) was refluxed with this isomer. (2)

    d Draw and name an isomer of pentanol that would react with acidified potassium dichromate(VI) but would not react with concentrated sulphuric acid. (2)

5 Glucose has the molecular formula $C_6H_{12}O_6$.

    a What is meant by the term molecular formula? (1)

    b What is the empirical formula of glucose? (1)

    c Glucose undergoes a reaction called fermentation. Describe the conditions required for this reaction and write an equation for the reaction. (1)

    d One important product of fermentation can be produced by direct hydration. Describe what advantage this process has over fermentation and write an equation for the reaction. (3)

6 a Epoxyethane is very reactive and is formed from ethene. Write an equation for this reaction and explain why epoxyethane is so reactive. (2)

    b Ethene reacts with concentrated sulphuric acid. Write a balanced equation for the reaction and name and illustrate the mechanism used. (4)

    c But-1-ene reacts with hydrogen bromide. Name the major organic product of this reaction and show the mechanism which forms this particular product. (4)

    d Why is the major product named in part **c** formed in preference to any other? (2)

7 a Why is fractional distillation commercially important for the petroleum industry? (3)

    b Why is cracking commercially important for the petroleum industry? (3)

    c Explain the main processes which go on in fractional distillation. (3)

    d Write an equation to demonstrate the two types of cracking. (2)

Answers

Answers to quick check questions

Module 1: Atomic structure, bonding and periodicity

page 3

1. In the nucleus

2. a number of protons

   b sum of number of protons and neutrons

   c atoms of an element with the same atomic number but different mass numbers

3. carbon-14

4.
	C	Ca	B	K	Al
P	6	20	5	19	13
E	6	20	5	19	13
N	6	20	6	20	14

page 5

1. vaporisation, ionisation, acceleration, deflection, detection

2. existence of isotopes

3. $\dfrac{(90.9 \times 20)+(0.26 \times 21)+(8.8 \times 22)}{(90.9+0.26+8.8)} = 20.18 = 20.2$ (1 d.p.)

page 7

1. a collection of sub-levels

2. Aufbau: Electrons always occupy the lowest energy sub-level first.
   Hund: Electrons occupy the orbitals first as unpaired electrons with same spin rather than as spin paired electrons.

3. Al $1s^2 2s^2 2p^6 3s^2 3p^1$
   Ca $1s^2 2s^2 2p^6 3s^2 3p^6 4s^2$

page 9

1. 3d

2.
a copper

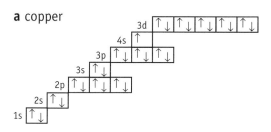

b silicon

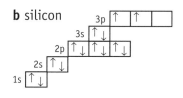

3. a Cu^+ $1s^2 2s^2 2p^6 3s^2 3p^6 3d^{10}$

   b Fe^{2+} $1s^2 2s^2 2p^6 3s^2 3p^6 3d^6$

page 11

1. the amount of energy that is required to remove one mole of electrons from one mole of atoms in the gaseous state, to form one mole of unipositive ions

2. 3 (3s 3p 3d)

3. The $3p^1$ electron is shielded from the nucleus by the complete 3s orbital and the singly occupied p orbital destabilises the electronic configuration of Al relative to the complete s orbital of Mg.

4. Chlorine has a larger effective nuclear charge than sulphur – hence higher first ionisation energy.

page 13

1. a $ZnCl_2$ = 136.4

   b NaOH = 40

   c H_2O_2 = 34

2. a calcium hydroxide = 74 g

   b copper(II) chloride = 134.5 g

   c ammonium nitrate = 80 g

3. a 9.8 g, b 35.5 g

4. a $CaCO_3$ = 0.1 mol, b NaOH = 0.25 mol,
   c $KMnO_4$ = 0.063 mol

5. $C_5H_{12}(g) + 8O_2(g) \rightarrow 5CO_2(g) + 6H_2O(g)$

Answers

page 15

1. The simplest whole number ratio of atoms of each element in a compound / The number of atoms of each element in a molecule of a compound.
2. The molecular formula is a multiple of the empirical formula.
3. C_2H_6O
4. $CClF_3$

page 17

1. Answer **a**
2. 5.85 g NaCl
3. 57.0 g

page 19

1. standard temperature and pressure (273 K and 100 kPa)
2. 22.4 dm^3
3. 16.8 dm^3
4. 30.4 dm^3

page 21

1. **a** 10 M, **b** 2 M
2. 50 cm^3

page 23

1.

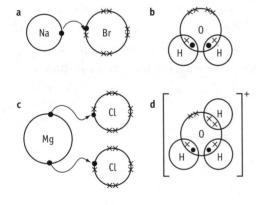

2. **a** ionic **b** covalent **c** ionic
   d covalent/co-ordinate (dative)
   e covalent **f** covalent

3.

page 25

1. Electronegativity is the ability of an atom to attract a pair of electrons in a covalent bond.
2. Down group: Increase in atomic radius and degree of shielding leads to a decrease in electronegativity.
   Across a period: Decrease in atomic radius and increase in nuclear charge leads to an increase in electronegativity.
3. The positive cations can attract the electron cloud on the anion and distort it especially if the cation is small and highly charged and the anion is large.

4.

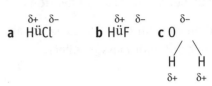

5. The difference in electronegativity of sodium and chlorine is much greater than that between silicon and chlorine.

page 27

1. **van der Waals forces**: These arise from the fluctuating movement of electrons, which causes temporary dipoles. These in turn induce attraction between molecules.
   Permanent dipole–dipole interactions: These arise when the electronegativities of the elements are very different, resulting in a polar bond.
   Hydrogen bonding: Nitrogen, oxygen and fluorine when bonded with hydrogen cause the bond with hydrogen to be so polarised that the hydrogen forms a weak bond with the nitrogen, oxygen or fluorine of another molecule.
2. The molecules have the same mass but those of pentane are more linear so have a greater surface area for van der Waals forces to act upon, hence the higher boiling point.
3. presence of hydrogen bonding in water

page 29

1.

	Kinetic energy of particles	Distance between particles
S	low	low
L	medium	low
G	high	high

2 The fact that gas particles are in constant rapid and random motion means that they will move and diffuse.

3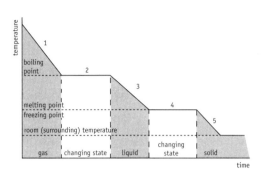

page 31

1 giant ionic lattices, giant metallic crystals, simple molecular structures, giant molecular crystals

2 availability of free electrons in metals

3 Graphite has a layered structure and the layers may slide over one another acting like a lubricant.

4 Only weak van der Waals forces hold them together. Only a little energy is needed to overcome these weak forces.

5 lithium

page 33

1 Lone pairs repel more strongly than bond pairs.

2 90°

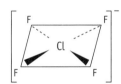

3 tetrahedral

4 a trigonal planar and trigonal pyramidal
  b Bond angle of NH_3 would increase to ~109° and that of BF_3 would decrease to ~109°.

page 35

1 alkali metals, alkaline earth metals, halogens, noble gases

2 a covalent bonding and giant atomic structure for the element b XCl_4

3 Atomic mass is not a periodic function like atomic number. It does not always increase from one element to the next (e.g. Te/I).

4 Generally the further electrons are from the nucleus the greater the shielding and the lower the first ionisation energy. The smaller the atomic radius the greater the attraction between the nucleus and the bond electrons as the shielding effect will be less and so the effective nuclear charge is larger and the electronegativity will be larger.

page 37

1 atomic radius and nuclear charge

2 Sodium to aluminium: Elements have giant metallic structures and the number of free electrons increases, increasing the strength of the metallic bond, so melting and boiling points increase.

Silicon has a giant molecular structure, with many strong covalent bonds so has a high melting and boiling point.

Phosphorus to argon: The elements have simple molecular structures with only van der Waals bonding, so melting and boiling points are low.

3 atomic radius ↓, first ionisation energy ↑, conductivity ↑ then ↓, electronegativity ↑, and melting and boiling point ↑ then ↓ (where ↓ = decrease and ↑ = increase).

page 39

1 atomic radius ↑, first ionisation energy ↓, electronegativity ↓, and melting point ↓ (where ↓ = decrease and ↑ = increase).

page 41

1 calcium hydroxide

2 Its hydroxide is amphoteric in nature; it bonds covalently; its maximum co-ordination number is 4.

3 It is covalent in nature and so soluble in organic solvents.

Module 2: Foundation physical and inorganic chemistry

page 45

1 1313.8 kJ mol^{-1}

2 anything sensible concerning heat loss to surroundings

3 54.6 kJ mol^{-1}

page 47

1 The first law of thermodynamics states that energy can't be created or destroyed – only changed from one form to another.

Answers

2 The enthalpy change in a reaction depends only on the initial and final states and is independent of the route taken.

3 −2548 kJ mol^{-1}

4 +2.5 kJ

page 49

1 Positive values mean that breaking the bond is endothermic as energy is used to separate species bonded together.

2 −93 kJ mol^{-1}

3 Mean bond enthalpies are average values for bond enthalpies, as they consider particular bonds in many different compounds.

4 The C=O bonds are in different environments and so will take differing amounts of energy to break.

page 51

1 Collisions may not have either the correct amount of energy to overcome the activation energy or the correct orientation of the molecules. There may not be sufficient collisions to produce a noticeable reaction.

2 A higher temperature will produce a greater rate of reaction as there is a greater number of particles with energy greater than or equal to the activation energy.

3 temperature, concentration, catalysis, pressure, surface area.

4 total number of collisions, fraction of collisions in correct orientation, fraction of particles colliding with energy $\geq E_a$.

page 53

1 When the rate of the forward reaction equals the rate of the reverse reaction, the reaction is in a **dynamic equilibrium**. If the phase of the reactants is the same it is homogeneous.

2 a moves the equilibrium position towards the reactants **b** reduces the rate

page 55

1 The products are made more economically.

2 Huge pressures would increase the yield but would be unworkable and uneconomic, so a lower pressure is used. Hence it is termed a compromise.

3 The rate of reaction would be too low.

4 Plant design may be expensive and working conditions may become dangerous if very high pressures are used. Low temperatures cause a slow reaction.

page 58

1

Oxidation is...	Reduction is...
addition of oxygen	loss of oxygen
loss of hydrogen	addition of hydrogen
loss of electrons	addition of electrons
an increase in oxidation state	a decrease in oxidation state

2 One causes oxidation and the other causes reduction.

3 a P_2O_5, **b** $HClO_3$

4 a Al = +3, **b** H = +1, **c** H = −1, **d** Cl = +3

5 a −3, **b** +2, **c** +1, **d** +4

page 61

1 $IO_3^- + 6H^+ + 5I^- \rightarrow 3I_2 + 3H_2O$ (oxidation/reduction)

IO_3^- oxidises I^-
I^- reduces IO_3^-

2 $Br_2(aq) + KI(aq) \rightarrow KBr(aq) + I_2(aq)$ (oxidation/reduction)

Br_2 oxidises I^-
I^- reduces Br_2

3 $ClO_3^-(aq) + 6H^+(aq) + 6e^- \rightarrow Cl^-(aq) + 3H_2O(l)$

4 $I_2(aq) + 2S_2O_3^{2-}(aq) \rightarrow 2I^-(aq) + S_4O_6^{2-}(aq)$

page 63

1 All require one electron to attain a full p sub-level.

2 More principal energy levels are needed, as the group is descended and...the outer principal energy level is increasingly further from the nucleus due to increased shielding ∴ larger

3 It is larger and the nuclear charge is shielded more by complete levels, so covalent bonding electrons are attracted less.

4 The Cl_2 molecules are smaller and so will have less van der Waals forces to overcome.

5 They are more likely to give up electrons than bromide ions as the nuclear charge is shielded more by complete energy levels.

Answers

page 65

1 They cause reduction by effectively donating electrons.

2 a $NaCl(s) + H_2SO_4(aq) \rightarrow NaHSO_4(aq) + HCl(g)$

   b $2NaBr(aq) + H_2SO_4(aq) \rightarrow Br_2(l) + SO_2(g) + 2H_2O(l)$

3 misty fumes = HI, black ppt = I_2, bad eggs smell = H_2S, yellow ppt = S

4 Use acidified silver nitrate solution – the colour of the ppt identifies the halide ion present. Ammonia solution and conc. ammonia can confirm results.

page 67

1 Green

2 One species is oxidised and reduced in the same reaction.

3 $Cl_2(g) + 2NaOH(aq) \rightarrow NaCl(aq) + NaClO(aq) + H_2O(l)$ or
   $Cl_2(g) + 2OH^-(aq) \rightarrow Cl^-(aq) + ClO^-(aq) + H_2O(l)$

4 A known amount of bleach is made into a standard solution.

   To a known sample of this standard solution, a known amount of solid potassium iodide is added with some sulphuric acid.

   The reactions are:

   $2H^+(aq) + Cl^-(aq) + ClO^-(aq) \rightleftharpoons Cl_2(g) + H_2O(l)$

   The chlorine produced displaces the iodide from KI and liberates iodine

   $Cl_2(aq) + 2I^-(aq) \rightarrow 2Cl^-(aq) + I_2(s)$

   This mixture is then titrated against a standard solution of sodium thiosulphate. A starch indicator detects the end point by going from blue/black to colourless.

   $I_2(aq) + 2S_2O_3^{2-}(aq) \rightarrow 2I^-(aq) + S_4O_6^{2-}(aq)$

page 69

1 iron ore (haematite); coke; limestone; hot air

2 $C(s) + O_2(g) \rightarrow CO_2(g)$

   $C(s) + CO_2(g) \rightarrow 2CO(g)$

   $Fe_2O_3(s) + 3CO(g) \rightarrow 2Fe(l) + 3CO_2(g)$

   $Fe_2O_3(s) + 3C(s) \rightarrow 2Fe(l) + 3CO(g)$

3 Limestone removes acidic oxide impurities like silica

   $CaCO_3(s) \rightarrow CaO(s) + CO_2(g)$

   $CaO(s) + SiO_2(s) \rightarrow CaSiO_3(l)$

4 Impurities in pig iron like carbon, sulphur and phosphorus are removed using the basic oxygen converter. Pure oxygen is blown through molten iron (usually with scrap metal in as well) to remove the carbon and sulphur impurities. Magnesium is also added to the molten iron to help remove the sulphur.

5
- Very high temperatures are required, which can be uneconomic and impractical.
- It cannot reduce the ores of titanium and tungsten, as these metals form carbides.
- Sulphide ores reduced by carbon sometimes cause atmospheric pollution in the form of sulphur dioxide.
- Carbon reduction also produces carbon monoxide which is poisonous and carbon dioxide which contributes to the greenhouse effect.

page 71

1 at the cathode   at the anode

   $Al^{3+} + 3e^- \rightarrow Al$ $2O^{2-} \rightarrow O_2 + 4e^-$

   the balanced electrode reactions

   $4Al^{3+} + 12e^- \rightarrow 4Al$

   $6O^{2-} \rightarrow 3O_2 + 12e^-$

2 to stop oxidation of Ti metal

3 the extraction process is expensive

4 lower impact on finite resources/reduces potential for pollution

Module 3: Introduction to organic chemistry

page 76

1 | Homologous series | General formula |
   |---|---|
   | alkanes | C_nH_{2n+2} |
   | alkenes | C_nH_{2n} |
   | haloalkanes | $C_nH_{2n+1}X$ (X is a halogen) |
   | alcohols | $C_nH_{2n+2}O$ |
   | aldehydes | $C_nH_{2n}O$ |
   | ketones | $C_nH_{2n}O$ |
   | carboxylic acids | $C_nH_{2n}O_2$ |

2 a pentane   b 2-methylpropane
   c 3-methylpentane   d 2-bromopropane

3 1-bromo-1-chloro-2,2,2-trifluoroethane

Answers

page 79

1. In both forms of isomerism the molecular formula and the functional group(s) are the same, but in chain isomers the arrangement of the carbon atoms in the chain (i.e. the carbon skeleton) is different and in position isomers the position of the functional group is different.

2. Three: pentane, 2-methylbutane, 2,2-dimethylpropane

3.
   1) C-C-C-C-C-C-C
      heptane
   2) C-C-C-C-C-C
          |
          C
      2-methylhexane
   3) C-C-C-C-C-C
            |
            C
      3-methylhexane
   4) C-C-C-C-C
           |
           C
           |
           C
      2,2-dimethylpentane
   5) C-C-C-C-C
           |
           C
      3,3-dimethylpentane
   6) C-C-C-C-C
         | |
         C C
      2,4-dimethylpentane
   7) C-C-C-C-C
         | |
         C C
      2,3-dimethylpentane
   8) C-C-C-C
        | |
        C C
        |
        C
      2,2,3-trimethylbutane
   9) C-C-C-C-C
          |
          C
          |
          C
      3-ethylpentane

4. The double bond does not allow free rotation along its axis, so the groups joined to the carbons making up the bond can be arranged differently giving *cis* and *trans* isomers.

5. functional group isomers

page 81

1. crude oil
2. jet fuel
3. any example from lower down the column; because the molecules are heavier and have higher boiling points
4. to prevent oxidation
5. to produce a larger supply of the more valuable smaller molecules

page 83

1. free radical substitution
2. initiation, propagation, termination
3. the Cl–Cl bond
4. Many products are possible as the reaction is a chain reaction.

page 85

1. See facts you need to know about alkenes on page 84
2. electrophilic addition
3. an electron-seeking species

4.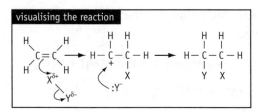

5. The second margarine has a higher degree of unsaturation as it takes more bromine water before enough was added to show the brown bromine colour.

page 87

1. Because the δ+ charge is effectively stabilised by a number of elctron-releasing groups, not just one

page 89

1. $C_3H_6(g) + H_2O(g) \rightarrow C_3H_7OH(g)$
   The major product is propan-2-ol as this goes via a secondary carbocation which is more stable than the primary carbocation which produces propan-1-ol.

2. $n(CH_2=CHCl) \rightarrow (CH_2-CHCl)_n$

3. The three-membered ring is strained.

page 91

1. a butan-2-ol   b butane-2-nitrile
   c 2-aminobutane   d but-1-ene and *cis*- and *trans*-but-2-ene

2.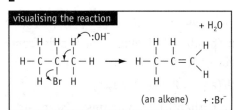

page 94

1. Fermentation:
   $C_6H_{12}O_6 \rightarrow 2C_2H_5OH + 2CO_2$
   Hydration of alkenes:
   $C_2H_4(g) + H_2O(g) \rightleftharpoons C_2H_5OH(g)$

2. Using Tollens reagent or Fehling's solution (see page 93 for details)

3. **1 protonation**
   Concentrated sulphuric acid protonates the hydroxyl group.
   2 loss of –OH$_2$
   The leaving group, –OH$_2$ is lost.
   3 loss of hydrogen
   Hydrogen is then lost from the carbocation to form the alkene.

4.

Answers to end of module exam style questions
Module 1: Atomic structure, bonding and periodicity

page 42

1 Electronic configuration describes the electron arrangement around the nucleus. It describes the position and orbital which an electron is in and it also tells us something about the relative energy of a particular electron.

2 Oxygen has one more e⁻ than nitrogen, which is spin paired in a p sub-level. This tends to destabilise the electronic configuration, so it is easier to remove.

3 Answer **a**. Mg has a larger IE than Na and Li as it has a greater nuclear charge, and Al has one p electron which is easier to remove than the spin paired s electron of Mg.

4 It represents the spin of the electron.

5 $V = 5.6$ dm^3

6 Empirical formula BH_3 and molecular formula B_2H_6

7 H_2O bond angle is 105°. O is in group VI, not an ion and has two atoms joined to it so has 8 e⁻ in outer shell, 2 bond pairs and 2 lone pairs which push the bond pairs together. H_3O^+ an extra bonding pair = pyramidal

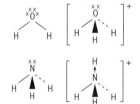

NH$_3$ bond angle is 107°. N is in group V, not an ion and has three atoms joined to it so has 8 e⁻ in outer shell, 3 bond pairs and 1 lone pair which pushes the bond pairs together. NH_4^+ has an extra bonding pair = tetrahedral (109.5°) In both cases the bond angle increases because there is less repulsion due to the extra bond pair.

8 a 612.2 moles
  b $2NaOH(aq) + H_2SO_4(aq) \rightarrow Na_2SO_4(aq) + H_2O(l)$
  c 1224.4 moles
  d 306.1 dm^3
  e If too much was added the lake would have an alkaline pH which could be as harmful as a low pH.
  f Use calcium carbonate.

9 a 3 moles of NaOH = 120 g
  b 1.5 moles of CaO = 84 g
  c 1.5 moles of Ca(OH)$_2$ = 111 g
  d 1.5 moles of Na$_2$CO$_3$ = 159 g

10 a
   b 28.0 g

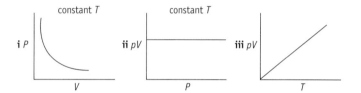

11 a **i** Vaporisation: the sample is vaporised by heating it electrically. It is then passed into the ionisation chamber.

ii Ionisation: the sample is then bombarded with high-energy electrons. This causes it to lose electrons and ionise, producing unipositive ions.

iii Acceleration: an electric field attracts the ions and accelerates them towards the detector. They all leave the ionisation chamber with similar velocities. This ion stream contains ions of different isotopic masses.

iv Deflection: the ion stream travels through a magnetic field. The ions are deflected by the field and as the heavier ions have more momentum than the lighter ions they are deflected less. This sorts out the mixture of ions in the ion stream into streams of ions each with a single isotopic mass.

v Detection: as the magnetic field is gradually increased each stream of isotopes hits the detector one at a time. The detector converts the information into a mass spectrum.

b 28.1

Answers

Module 2: Foundation physical and inorganic chemistry

page 72

1. **a** The enthalpy change when 1 mole of compound is formed from its elements in their standard states under standard conditions.
   b -0.4 kJ mol^{-1}

2. **a** reactions iii and iv **b** reaction iii
   c reactions i and ii **d** reaction iv

3. methanol = 710.98 kJ mol^{-1}
   ethanol = 1350.47 kJ mol^{-1}
   propan-1-ol = 2013.48 kJ mol^{-1}
   butanol ~ 2675 kJ mol^{-1}

4. **a** A = enthalpy of the reaction
   B = activation energy for uncatalysed reaction
   C = activation energy for catalysed reaction
   b the minimum amount of energy a reactant requires to cause a reaction on collision
   c kinetic energy ($\frac{1}{2}mv^2$) ∝ temperature; a small temperature rise causes a large increase in molecular energies, increasing the chance of a collision and making more of the collisions effective, so speeding up the reaction.
   d The catalysed reaction goes via a different reaction pathway which has a lower E_a. So the number of particles with $E \geq E_a$ increases, so more particles have sufficient energy to react.

5. **a** $Cl_2(g) + H_2O(l) \rightleftharpoons HCl(aq) + HClO(aq)$
   b green
   c i Green fades as equilibrium moves to RHS.
   ii $Cl_2(g) + 2NaOH(aq) \rightarrow NaCl(aq) + NaClO(aq) + H_2O(l)$
   iii Green colour returns.
   $HCl(aq) + HClO(aq) \rightarrow Cl_2(g) + H_2O(l)$
   d NaClO
   e +1

6. **a** Solid: the trend is for increasing mass and increasing size, so there will be a larger surface for VDW forces to act upon.
   b i oxidising agent
   ii +6
   iii −2
   c Add acidified silver nitrate then dilute then conc. ammonia solution to the samples: The results below would allow identification.

halide / reactant	Cl^- (aq)	Br^- (aq)	I^- (aq)
$AgNO_3$(aq)	white ppt	Cream ppt	yellow ppt
Dilute NH_3(aq)	soluble	insoluble	insoluble
Concentrated NH_3(aq)	soluble	soluble	insoluble

7. **a** metal oxide + carbon
   e.g. $2Fe_2O_3(l) + 3C(s) \rightarrow 4Fe(l) + 3CO_2(g)$
   metal chloride + active metal
   e.g. $TiCl_4 + 4Na \rightarrow Ti + 4NaCl$
   electrolysis of molten ore
   e.g. $2Al_2O_3 \rightarrow 4Al + 3O_2$
   b metal oxide + carbon: cheap reductant; carbon abundant; continuous process; purity not an issue as carbon impurities can be dealt with.
   metal chloride + active metal: carbon reduction forms the carbide TiC; Ti is needed in pure form.
   electrolysis of molten ore: carbon can't reduce Al; temperature would have to be too high; usefulness of Al outweighs cost of producing it; cryolite lowers melting point so lowers energy requirements.

Module 3: Introduction to organic chemistry

page 95

1. **a** Mechanism: the mechanics of the reaction – which atoms attack, which bonds are broken and which bonds form
   b Electrophile: an electron-seeking or electron-deficient species
   c Nucleophile: species with at least one unshared pair of electrons
   d Free radical: a species which possesses an unpaired electron
   e Electrophilic addition: where an electrophile adds across a double bond

f Nucleophilic substitution: where a nucleophile replaces a species

g Elimination: where a group such as water is removed from the molecule

Functional group	Name of reaction mechanism commonly used
alkanes	free radical substitution
alkenes	electrophilic addition
haloalkanes	nucleophilic substitution and elimination
alcohols	elimination

3 a free radical substitution

b to break the Cl–Cl bond to initiate the reaction

c $CH_4 + Cl_2 \rightarrow HCl + CH_3Cl$

d $CHCl_2\bullet + Cl\bullet \rightarrow CHCl_3$ (trichloromethane)

e ethane C_2H_6

f In the propagation stage each free radical that reacts produces another free radical to continue the reaction. This results in a mixture of products in a chain reaction.

4 a $CH_3CH_2CHCH_2CH_3 + [O] \rightarrow CH_3CH_2CCH_2CH_3$
    $\quad\quad\quad\quad |\quad\quad\quad\quad\quad\quad\quad\quad\quad\quad ||$
    $\quad\quad\quad\quad OH\quad\quad\quad\quad\quad\quad\quad\quad\quad O$

b i dehydration, **ii** 180°C, conc. H_2SO_4
iii

c no change to orange colour

d

2,2-dimethylpropanol

5 a the formula that shows the actual number of each kind of atom present in a molecule

b CH_2O

c fermentation: $C_6H_{12}O_6 \rightarrow 2C_2H_5OH + 2CO_2$
conditions: 37°C, yeast, batched over several days

d Advantage: hydration is fast, produces pure ethanol and is continuous (which is cheaper).
$C_2H_4(g) + H_2O(g) \rightleftharpoons C_2H_5OH(g)$

6 a $2CH_2=CH_2 + O_2 \xrightarrow{air/Ag/180°C} 2H_2C-CH_2$ (epoxide)

It is reactive due to the very high strain on the ring.

b $CH_2=CH_2 + H_2SO_4 \rightarrow CH_3C(HSO_4)H_2$
electrophilic addition

c 2-bromobutane

d It is formed from a secondary carbocation, which is more stable than a primary one.

7 a Crude oil is split into fractions which are more useful; crude oil is of little use on its own; the fractions provide feedstock for the chemical and polymer industries.

b Hydrocarbons with large molecules are less useful than those with smaller ones; there is more demand for hydrocarbons with small molecules than for those with larger ones; cracking produces alkenes, which can be used to make polymers.

c Heated crude oil is passed into distillation column at ~500°C; the different boiling point fractions boil off; the column has a temperature gradient so fractions with lower molecular mass rise to the top and are cooled and condensed; use of low pressure can stop decomposition or cracking of the fractions; heavier fractions are taken off lower down the column.

d Any example where equation is balanced.

Index

abundance, relative 4, 5
acceleration of ions 4
acid rain 69, 82
activation energy (E_a) 48, 50, 51
addition reactions 84–5, 88
alcohols 74, 75, 78, 92–4
aldehydes 74, 75, 76, 93
alkanes 74, 75, 80–3
    haloalkanes 74, 75, 90–1
alkenes 74, 75, 84–9
    from alcohols 94
    from haloalkanes 91
    geometrical isomerism 79
    hydration 88, 92
    isomerism 79, 84, 86–7, 91
alkoxyalcohols 89
aluminium (Al) 30, 36–7
    extraction 70, 71
    oxidation state 57
aluminium chloride 23, 25
aluminium oxide 70
amines, from haloalkanes 91
ammonia 27, 32, 33
    manufacture 54
    reaction with haloalkanes 91
ammonium ions 23
amphotericity 40, 41
argon (Ar) 11
aromatic hydrocarbons 81
-ate ions 57, 58
atomic mass (A) 2–3
    relative (A_r; RAM) 4, 5, 12
atomic number (Z) 2–3, 34
atomic radius 35
    and electronegativity 24, 39
    of group II elements 38
    of group VII elements 62
    of Period 3 elements 11, 36
atoms
    electronic configuration 6–7
    number of 12
    oxidation states 57, 58
Aufbau principle 7
Avogadro constant 12
Avogadro's law 18–19

balanced equations 13, 21
barium (Ba) 38–9, 40–1
barium sulphate ($BaSO_4$) 40
basic oxygen converters 69
batch process 71
bauxite 70
beryllium (Be) 38–41
beryllium chloride ($BeCl_2$) 41
bitumen 80
blast furnaces 68–9
bleach 66
boiling 28–9
boiling points 29, 35
    of group VII elements 62
    of Period 3 elements 37
    of petroleum fractions 80
    and structure and bonding 37
bomb calorimeters 45

bond dissociation 48–9, 80, 83, 91
bond enthalpy
    and reaction rate 80, 83, 91
    standard molar 48–9
bond fission
    heterolytic 85
    homolytic 48, 80–1, 83
bond polarity 24
bonding
    in C=C double bonds 79
    covalent, see covalent bonding
    hydrogen 27
    intermolecular 26–7, 31, 37
    intramolecular 22–5, 37
    ionic 22, 25
    and physical properties 37
    polar 24, 27, 90
    in solids, liquids and gases 28
    and structure 30–1, 37
    van der Waals 26, 31, 37, 80
Boyle's law 19
bromide ions (Br^-) 64–5
bromine (Br) 62–3, 84, 86–7
bromoethane 90–1
bromopropane 86–7
butanal 78
butane 14, 80
butanol 74, 76
    isomerism 78, 92
    reaction with H_2SO_4 94
butanone 78
butene 84
    from butanol 94
    isomerism 79, 84, 91

C–C bonds 80–1
C=C double bonds 79, 84–5
C–H bonds 49, 80
C–halogen bonds 91
caesium chloride (CsCl) 30
calcium (Ca) 8, 38–40
calorimetry 45
carbocations
    in alkene reactions 86–7
    in catalytic cracking 81
    in dehydration of alcohol 94
    in electrophilic addition 85
    stability 87
carbon (C)
    from combustion 81
    reduction by 68–9, 70, 71
carbon dioxide (CO_2) 33
    from combustion 81
    from metal extraction 68, 69, 70
carbon monoxide
    from combustion 81, 82
    from metal extraction 68, 69, 70
carboxylic acids 74, 75, 76,
    isomerism 78
cast iron (pig iron) 68, 69
catalysts
    in Contact process 55
    for cracking 81

and E_a 50, 51
and equilibria 53
in Haber process 54
for hydration of alkenes 88
for hydrogenation 88, 94
and reaction rate 51
catalytic converters 82
chain isomerism 77, 78
chain reactions 83
changes of state 28–9, 35
    see also boiling points; melting points
charge, nuclear 10, 11, 24, 36
charge density 25
charge separation in bonds 24
Charles' law 19
chemical properties
    of geometrical isomers 79
    of group II elements 38, 40–1
    of group VII elements 63
    of isotopes 3
    of metals and non-metals 34
chlorate(I) (ClO^-) ions 66–7
chlorate(V) (ClO_3^-) ions 58
chloride (Cl^-) ions 9, 64, 65
chlorination of alkanes 83
chlorine (Cl) 36–7, 62–3, 66–7
    dot and cross diagram 23
    ionisation energy 11, 36
    isotopes 3
    oxidation of Fe^{2+} 61
    oxidation states 57, 58
    as oxidising agent 59, 61, 63
    reaction with alkanes 83
    reaction with KI 59
    reaction with methane 83
    relative atomic mass 5
chloroethane 75
chloromethane 83, 90
chromium (Cr) 8, 9
chromium(III) ions 9
cis 79, 84
cis-trans isomerism 77, 79, 84
closed systems, equilibrium in 52
co-ordinate bonds 23, 41
co-ordination number 41
collision theory 50–1
combustion 44, 81, 82, 92
    standard enthalpy change ($\Delta H_c^\ominus$) 44–5, 46, 47
complex ions 41, 57
compounds
    amphoteric 40, 41
    ionic 22
    organic 74–6, 84
    oxidation states 57–8
concentration 20–1
    at dynamic equilibrium 52
    and Le Chatelier's principle 53
    and reaction rate 51
    from titrations 21
condensation 29
conductivity of Period 3 elements 36

Contact process 55
cooling curves 29
copper (Cu) 8, 30
covalent bonding 23, 25
    double 23, 33, 79
    in macromolecular crystals 31
    and oxidation states 57
cracking of alkanes 80–1
crude oil 80–1
cycles of reactions 46–7
cyclohexane test for halogens 63

d block elements 7, 34
d electron sub-levels 6–7, 8–9
dative covalent bonds 23
deflection of ions 4–5
dehydration of alcohols 94
density 45
detection of ions 4
deuterium (D) 3
diamond 31
dibromopropane 90
dibromopropanoic acid 76
dichloropentane 78
diesel engines 82
diesel oil 80
dimers and dimerisation 23
dipoles 24, 26, 27, 35, 85
direct hydration 88
direct reactions 46–7
displacement reactions
    enthalpy change 45
    of halogens 63
    in metal extraction 68, 70–1
distillation, fractional 80
dot and cross diagrams 22, 23
double covalent bonds 23, 33, 79, 84–5
dynamic chemical equilibrium 52

economics
    of ethanol production 92
    of Haber process 54
    of metal extraction 70, 71
electrical conductivity 36
electrolysis, to extract metals 68, 70
electron deficiency
    and bond polarity 24
    in electrophiles 84–5
electron pair repulsion theory 32
electron-releasing groups 87
electron rich atoms 24
electronegativity 24, 35
    and atomic radius 24, 39
    and electronic shielding 39
    of group II elements 39
    of group VII elements 62
    of Period 3 elements 36
electronic configuration
    of atoms 6–7, 8–9, 10, 36, 38
    and ionisation energy 10
    of ions 8–9, 10
electronic energy levels 6–7, 8–9, 10–11

Index

electronic shielding
   and atomic radius 36, 38–9
   and electronegativity 39
   and ionisation energy 10–11, 36, 38–9
electrons 2–3
   bond pairs 32, 33, 85
   delocalised 30
   lone pairs 32–3, 85, 90–1
   movement in equations 85
   shared in covalent bonds 23
   shells and sub-shells 6–9, 10–11
   transfer in ionic bonds 22
electrophiles 84, 87
electrophilic addition 84–5
electrostatic forces
   in ionic bonding 22, 30
   in polar bonding 27
elements 2
   electronic configuration 8
   periodicity of properties 34–5
   with variable oxidation states 57–8
elimination reactions 91, 94
empirical formulae 14–15, 74
endothermic reactions 44, 48, 52
energy, kinetic 50
energy levels of electrons 6–7, 8–9, 10–11
enthalpy (H) 44, 52
enthalpy change (ΔH) 44, 45, 46–9
   of bond dissociation 48–9
   of combustion ($\Delta H_c^\ominus$) 44–5, 46, 47
   of formation ($\Delta H_f^\ominus$) 44, 46, 47
   of fusion 28
   standard ($\Delta H^\ominus$) 44
   of vaporisation 29
epoxyethane 88–9
equations 13
   electron movement in 85
equilibria 52–3
esters 78
ethanal 75, 93
ethane 75, 80
ethane-1,2-diol 89
ethanoic acid 75, 93
ethanol 74, 75, 88, 90, 92
ethene 74, 75, 84
   hydration 88, 92
ethers 78
   epoxyethane 88–9
ethylamine 91
evaporation 29
exothermic reactions 44, 48, 49
   in Haber process 54
   and Le Chatelier's principle 52
explosive reactions 83
extraction of metals 68–71
extrapolation of graphs 45

Fehling's solution 93
fermentation of sugars 92
first ionisation energy 10, 36, 38–9
first law of thermodynamics 46
fission see bond fission
fluorine (F) 27, 57, 62
fractional distillation 80
free radicals 80–1, 83

freezing (solidification) 29
fuel oil 80
fuels 80, 81, 82, 92
functional groups 74, 75, 76
   isomerism 77, 78
fusion, enthalpy change 28

gases 28
   changes of state 28–9
   kinetic theory 19
   volume calculations 18–19
gasoline (petrol) 80
general formulae 74, 75
geometrical isomerism 77, 79, 84
giant structures 22, 30, 31, 37
graphite 31
groups in Periodic Table 34, 38
   group II 38–41
   group VII see chlorine; halide ions; halogens
   oxidation states 57

Haber process 54
haematite 68
half equations 13
   for redox reactions 56, 59–61
halide ions 64–5
haloalkanes (halogenoalkanes) 74, 75, 90–1
halogens 57, 62–3
   bromine 62–3, 84, 86–7
   chlorine see chlorine
   fluorine 27, 57
   iodine 31, 62, 63, 67
Hess's law 46–7
heterogeneous equilibria 52
heterolytic fission 85
hexane 80
hexene 84
highest occupied energy level 9
homogeneous equilibria 52
   in Contact process 55
   in Haber process 54
homologous series 74, 75, 84
   isomerism 78
homolytic fission 48, 80–1, 83
Hund's rule 7
hydration of alkenes 88, 92
hydrides 57
hydrocarbons
   aromatic 81
   unburned 82
   see also alkanes; alkenes
hydrogen (H) 3, 7, 57, 80
hydrogen bonding 27
hydrogen bromide 86–7
hydrogen chloride 27, 53
hydrolysis 89, 90

ICl_4^- 32
ideal gas equation 19
indirect reactions 46–7
induced dipoles 85
inductive effect 87
industrial processes
   Contact process 55
   ethanol production 88, 92
   extraction of metals 68–71
   Haber process 54
   hydration of alkenes 88
   hydrogenation of alkenes 88
   polymerisation of alkenes 88

infixes in chemical names 76
initiation of free radical substitution 83
intermediates 83
intermolecular forces 26–7, 31, 37
internal combustion engines 82
iodate(V) ions (IO_3^-) 60
iodide ions 64, 65
iodine (I) 31, 62, 63
   reaction with $Na_2S_2O_3$ 67
iodoethane 90
ion–electron equations 13
   for redox reactions 56, 59–61
ionic bonding 22, 25
ionic equations 13, 56
ionic lattices, giant 22, 30
ionisation 4
ionisation energy 10–11, 35
   first 10, 36, 38–9
   of group II elements 40
ions
   electronic configuration 8–9
   halide 64–5
   in mass spectrometry 4, 5
   oxidation states 57, 58
   polarisation 25
   size 9
   of transition metals 8, 9
iron (Fe)
   as catalyst 54
   electronic configuration 8
   extraction 68–9, 71
   oxidation states 58
   structure and properties 30
iron(II) ions, oxidation 61
isoelectronic ions 9
isomerism 77–9
   of alkenes 79, 84, 86–7, 91
isotopes 3
   separation 4, 5
IUPAC rules for naming compounds 76

kerosene 80
ketones 74, 75, 93
kinetic energy, and temperature 50
kinetic theory 19, 28–9
kinetics 50–1
   and bond enthalpy 91
   in Haber process 54

Le Chatelier's principle 52–3
Lewis acid, in catalytic cracking 81
lime (CaO) 69
limestone 68–9
linear molecules 32
liquids 28
   changes of state 28–9, 35
lithium (Li) 3
lone pairs of electrons 32–3
   movement 85, 90–1
   in nucleophiles 90–1
lowest occupied energy level 9

macromolecular crystals 31
macroscopic properties, at dynamic equilibrium 52
magnesium (Mg)
   atomic radius 11, 36
   chemical properties 38, 40–1

co-ordination number 41
   ionisation energy 11, 36
   physical properties 30, 36–7, 38–9
   as reducing agent 71
magnesium carbonate, M_r 12
magnesium chloride 25
magnesium oxide 30
manganese (Mn) 58
margarine 88
mass
   conversion to moles 13
   reactant and product calculations 16–17
mass number (A) 2–5
mass spectra 3, 4–5
mass spectrometry 4–5
mass to charge (m/z) ratio 5
Maxwell–Boltzmann distribution curves 50
mean bond enthalpy 49
melting (fusion) 28
melting points 29, 35
   of group II elements 39
   of Period 3 elements 37
   and structure and bonding 37, 39
metal oxides, reduction 68–71
metallic bonding 30, 39
metallic structures, giant 30
metals
   extraction 68–71
   group II elements 38–41
   in Periodic Table 34
methane 32, 80
   $\Delta H_c^\ominus$ 44, 47
   mean bond enthalpy 49
   reaction with chlorine 83
methanol 92
methyl (CH_3) group 87
methylpropanol 78, 92
mineral oil 80
molar bond enthalpy 48–9
molar gas volume 18–19
molarity 20–1
molecular crystals 31
molecular energy distribution 50
molecular equations 13
molecular formulae 14, 15, 74
molecular mass, relative (RMM; M_r) 4, 12–13
moles 12–13, 16–17, 18–19, 20–1
motor fuels 80, 81, 82

names
   for organic compounds 74–6, 84
   for variable oxidation states 57–8
naphtha 80
neutralisation reactions 45
neutrons 2–3
NF_2^+ ions 32
nickel (Ni) 88, 94
nitriles 90
nitrogen (N)
   hydrogen bonding 27
   polyatomic ions 58
nitrogen oxides (NO_x) 53, 82
nitrous oxide (NO) 82
nomenclature see names
non-metals 34
non-polar hydrocarbons 82
nuclear charge
   and electronegativity 24, 36

Index

and ionisation energy 10, 11, 36, 38–9
nucleon number (A) 2–5
nucleophiles (:Nu) 90
nucleophilic substitution 90–1
nucleus 2

O–H bonds, dissociation enthalpy 49
octahedral molecules 32
orbitals, electronic 6, 7
orientation, and reaction rate 50
oxidation 56
    of alcohols and aldehydes 93
    of Fe^{2+} 61
    of KI 59
oxidation states or numbers (Ox; ON) 56–8, 59, 60, 61
oxidising agents 56
    H_2SO_4 64–5
    halogens 63, 64
oxygen (O) 7, 27, 57

p block elements 7, 34
p electron sub-levels 6–7, 8–9, 10–11
particle model of matter 28–9
particles, fundamental 2–3
pentane 76
pentene 84
Period 3 elements 11, 36–7
Periodic Table 2, 34, 36
    diagram *inside front cover*, 34
    periodicity in 34–5, 36–7
permanent dipoles 24, 27
petrol (gasoline) 80
petroleum 80–1
phases of matter 28–9, 52
phosphoric acid 88
phosphorus (P) 11, 36–7
phosphorus(III) chloride 58
phosphorus(V) chloride 32
phosphorus(V) oxide 69
physical properties
    of geometrical isomers 79
    of group II elements 38–9
    of halogens 62
    of ionic compounds 22
    of isotopes 3
    of metals and non-metals 34
    of Period 3 elements 36–7
    of states of matter 28
    and structure 30–1
    and van der Waals forces 26
pi (π) bonds 79, 84
pig iron (cast iron) 68, 69
plasticisers 89
plastics, from alkenes 80, 88
plateaus on cooling curves 29
platinum (Pt) 88–9, 94
polar bonding 24, 27, 90
polarisation of ions 25
pollution 69, 81, 82
poly(alkoxyalcohol)s 89
polyatomic ions, oxidation state 58
poly(ethene) 80, 88
polymerisation of alkenes 88
poly(propene) 88
polyunsaturates 88
positional isomerism 77, 78

potassium dichromate(VI) 93
potassium iodide 59
potassium ions (K^+) 8, 9
potassium (K) 9, 10–11
prefixes in chemical names 76
pressure
    in Contact process 55
    in Haber process 54
    and Le Chatelier's principle 53
    and reaction rate 51
primary alcohols 92, 93
primary carbocations 87
principal energy levels of electrons 6–7
    and ionisation energy 10–11
propagation of free radical substitution 83
propane 80
propanoic acid 90
propanol 74, 92, 93, 94
propanone 75, 93
propene, 14, 74, 75, 84
    from bromopropane 91
    from propan-2-ol 94
proton number (Z) 2–3, 34
protons 2–3

radium (Ra) 38–9
reaction cycles 46–7
reaction rates (kinetics) 50–1
    and bond enthalpy 91
    in Haber process 54
reactions
    endothermic 44, 48, 52
    exothermic 44, 48, 49, 52, 54
    redox 56–7, 59–62
    reversible 52–3
recycling, of metals 71
redox reactions 56–7, 59–61, 66
reducing agents 56
    halide ions 64–5
    sodium tetrahydridoborate(III) 94
reduction 56
    of aldehydes and ketones 94
    of alkenes 88
    of Cl_2 59
    of metal oxides 68–71
refinery gases 80
reflux apparatus 93
relative abundance 4, 5
relative atomic mass (RAM; A_r) 12
relative charge 2
relative mass 2, 4–5
relative molecular mass (RMM; M_r) 4, 12–13
reversible reactions 52–3
rutile (TiO_2) 70–1

s block elements 7, 34
s electron sub-levels 6–7, 8–9, 10–11
saturated hydrocarbons 82, 88
secondary alcohols 92, 93
secondary carbocations 87
semi-conductors, silicon 36
shapes
    of alkenes 84
    of group II complexes 41
    of molecules 32–3
    of orbitals 6

shielding by electrons
    and atomic radius 36, 38–9
    and electronegativity 39
    and ionisation energy 10–11, 36, 38–9
silicon (Si) 11, 36–7
silicon tetrachloride 25
silver nitrate test for halide ions 65
size
    of atoms *see* atomic radius
    of ions 9
slag 68–9
sodium (Na) 11, 36–7, 71
sodium bromide 64
sodium carbonate, M_r 12
sodium chlorate(I) (NaClO) 66
sodium chloride 25, 30
    dot and cross diagram 22
    M_r 12
    reaction with H_2SO_4 64
sodium hydroxide 66, 90
sodium iodide 65
sodium tetrahydridoborate(III) 94
sodium thiosulphate 66–7
solidification (freezing) 29
solids 28
    changes of state 28, 29, 35
solubility of group II compounds 40–1
specific heat capacity 45
spin pairing of electrons 7
standard enthalpy change ($\Delta H^\ominus$) 44
    see also enthalpy change
standard temperature and pressure (stp) 18
states of matter 28–9
    in equilibria 52
steel 69, 71
stereoisomerism 77, 79
strontium (Sr) 38–9, 40
structural formulae 75
structural isomerism 77
structures
    giant 22, 30, 31, 37
    and physical properties 30–1, 37
sub-atomic particles 2–3
sub-shells (sub-levels) of electrons 6–9, 10–11
    in ionic bonding 22
sublimation 28
substitution reactions
    of functional groups 74
    nucleophilic 90–1
suffixes in chemical names 76
sugar, fermentation 92
sulphate ions (SO_4^{2-}) 33
sulphur (S) 11, 36–7, 58
sulphur dioxide 69
sulphur trioxide 33, 55
sulphuric acid, concentrated
    manufacture 55
    reactions 64, 86, 87, 94
surface area
    and reaction rate 51
    and van der Waals forces 26
surfactants 89

tar 80

temperature
    in Contact process 55
    in Haber process 54
    and kinetic energy 50
    and Le Chatelier's principle 52
    and particle energy for evaporation 29
    and reaction rate 51
temporary dipoles 26
termination of free radical substitution 83
tertiary alcohols 92
tertiary carbocations 87
tests
    for aldehydes and ketones 93
    for alkenes 84
    estimation of ClO^- 66–7
    for halide ions 65
    for halogens 63
    for sulphate 40–1
tetrahedral molecules 32, 33
thermal cracking 80–1
thermodynamics, first law of 46
titanium (Ti) 69, 70–1
titanium(IV) carbide 69, 71
titanium(IV) chloride 71
titanium(IV) oxide 70–1
titrations
    calculations for 21
    for ClO^- estimation 66–7
Tollens reagent 93
trans 79, 84
transition metal ions 8, 9
translational energy 28
trigonal planar molecules 32
trigonal pyramidal molecules 32, 33
triple bonds 23, 33
tritium (T) 3
tungsten (W) 69

units
    for ideal gas equation 19
    see also moles
unsaturated hydrocarbons 88

van der Waals forces (VDWs) 26, 31, 37, 80
vanadium(V) oxide 55
vaporisation
    enthalpy change of 29
    in mass spectrometry 4
volume calculations
    from titrations 21
    for gaseous reactants and products 18–19

water
    bond enthalpy 48–9
    density 45
    molecular shape 32, 33
    purification treatment 66
    reactions with group II elements 40–1

yield of reaction 52

zeolite catalyst 81
zinc (Zn) 30